Advances in Agricultural Biotechnology

Advances in Agricultural Biotechnology

Ramanand Sagar

CYBER TECH PUBLICATIONS
4264/3, Ansari Road, Daryaganj, New Delhi-110002 (India)
Ph.: 011-23244078. Fax: 011-23280028
E-Mail: cyberpublicationsdelhi@yahoo.com

Advances in Agricultural Biotechnology
Ramanand Sagar

First Edition 2009

Published by :

G.S. Rawat for **Cyber Tech Publications**
4264/3, Ansari Road, Daryaganj, New Delhi-110002 (India)
Ph.: 011-23244078. Fax: 011-23280028
E-Mail: cyberpublicationsdelhi@yahoo.com

Printed at : **Asian Offset Printers**

Preface

Biotechnology has long been used to make everyday products (e.g. the use of micro-organisms, such as bacteria or fungi, to manufacture cheese, wine, and antibiotics). The scope of this term has changed over time as new technologies have been invented. The term "biotechnology" came into wide-spread use in the 1970s following the development of genetic engineering. In looking back, however, it is evident that humans have used "biotechnology" in its simpler forms for thousands of years. For example, civilizations have actively developed and selected superior plants and livestock animals. The scientists broadly defines biotechnology as: the application of science and engineering to the direct or indirect use of living organisms or parts or products of living organisms in their natural and modified forms.

Now the world has a unique regulatory approach for biotechnology products that considers the novelty of a product, not its method of production, as the trigger for regulatory review. For example, a new agricultural product may be considered novel if it has one or more new traits or one or more changed traits, or if it has a new use. The criteria for novelty are specific to the product (e.g. novel livestock feeds, novel fertilizers and supplements, and plants with novel traits are all defined in the appropriate Acts and Regulations). Novel agricultural products can be developed through various genetic modification techniques (traditional breeding, genetic engineering, mutagenesis, cell fusion, etc).

Regulatory groups and policy makers around the world define biotechnology differently. The differences tend to centre on how broadly the term should be defined. For example, some countries may limit the definition of biotechnology to include modern biotechnology techniques, such as genetic engineering, while others' definition includes other more traditional techniques.

Since the earliest days of agricultural production, people have developed new plant and animal breeds to suit their needs - perhaps to provide more food or to withstand particular environmental conditions. Micro-organisms have also played a role in food production. Today, the agriculture and agri-food sector has the same objectives, but many new tools are available to achieve them.

The undersigned is confident enough that this work of him would be most informative to the students and teachers alike, suggestions and complaints if any are always welcome.

—Author

Contents

Contents

1

Agricultural Biotechnology Research in India: Status and Policies

Agriculture is a way of life for more than sixty per cent of India's population. The cultivation of land not only sustains their livelihood but also provides a social milieu for their day-to-day living. No wonder the hopes, despairs, joys and sorrows of rural communities are woven around what the land provides. Around 35 years ago, agricultural production in India got a major boost with the introduction of dwarf varieties of wheat and rice.

The introduction of these varieties led to a dramatic increase in the yields of the two crops. Some productivity enhancement came through the use of hybrids in corn, sorghum and millet although the area under cultivation of these crops has steadily decreased in the last decade.

In the last 10 years the yields of rice and wheat have also plateaued out. The productive agricultural areas in the North, due to continuous rice–wheat cultivation are encountering serious problems of sub-soil water depletion, deficiency of micronutrients in the soil and increase in the use of pesticides, fungicides and herbicides to control pests, pathogens and persistent weeds.

Agricultural production is becoming more and more dependent on agro-chemicals, thereby increasing input costs and causing significant damage to the environment and human health. Both farmers and consumers are at the receiving end – farmers by exposure to agrochemicals and consumers due to residues of agrochemicals in the consumed food.

While there is self-sufficiency in cereal grains at present, the yields and productivity of dryland crops, mostly grain legumes and oilseeds remain low and no major breakthroughs in productivity enhancement and yield stabilization have been achieved. Currently, India is importing both grain legumes and edible oils to meet internal demands. India's population is expected to reach approximately 1.5 billion by 2050.

It is estimated that around 300 million (roughly 30%) of India's population is suffering from malnutrition. Due to socio-economic reasons, women and children are more at the receiving end in terms of micronutrient deficiency and low calorie intake. Nutritional security for everyone would require more extensive availability of grain legumes, edible oils, fruits and vegetables, milk and poultry products.

The challenges of malnutrition, enhanced productivity and crop diversification can be met by better resource management and by breeding more productive, more nutritious and at the same time less resource input demanding crops.

Crop biotechnology, which broadly includes areas of development of transgenic crops, structural and functional genomics and marker-assisted breeding could provide us with the vital breakthroughs to achieve improvements in both quality and quantity in a sustainable manner. With the advent of techniques of genetic engineering in the early seventies, the natural barrier to gene exchange has been removed. Sequences from varied sources like bacteria, viruses and eukaryotic systems can be transferred to plants to develop transgenic crop varieties. Achievements, to date, in plant biotechnology have surpassed all previous expectations and with the development of high throughput instruments, the future is even more promising.

In this article we outline the contributions made by the Department of Biotechnology in promoting research and development in the area of crop biotechnology and in establishing proper regulatory regimes for transgenic crops.

Recognizing the importance and potential of biotechnology, the Government of India set up the National Biotechnology Board in 1982 which became a full-fledged department (called as

Department of Biotechnology, DBT) in 1986 under the Ministry of Science and Technology. The plan-wise budget of DBT from 1992 onwards in the area of crop biotechnology is given in Table 1. To provide research and human resource development in crop biotechnology, DBT supported the establishment of Centres for Plant Molecular Biology (CPMB) in 1990. A total of six such centres were set up initially at various Universities/Institutions namely, Jawaharlal Nehru University (New Delhi), Madurai Kamaraj University (Madurai), Tamil Nadu Agricultural University (Coimbatore), Osmania University (Hyderabad), National Botanical Research Institute (Lucknow) and Bose Institute (Kolkata). A seventh centre was established at the University of Delhi South Campus in 1997. A total financial outlay of Rs 45.98 crores was made available for the development and research activities of CPMBs.

Most of the research on crop biotechnology currently underway in India is being supported by the Government and the DBT has been the main agency amongst government departments to support research endeavours in this area. The DBT has supported over the last 12 years a large number of research projects which deal with the development of *in vitro* regeneration and genetic transformation protocols of important crop species grown in India, the development of transgenics with genes of agronomic importance and several projects dealing with basic research in plant molecular biology.

To bring investigators with expertise in different facets of biology together to address defined research goals, a number of multi-institutional projects have been launched which include development of transgenics for resistance to geminiviruses in cotton, mungbean and tomato, resistance to rice tungro disease, resistance to bollworms in cotton, development of nutritionally improved potato with a balanced amino acid composition using a heterologous protein from *Amaranthus* and development of molecular methods for heterosis breeding. Another set of projects include tagging of quality traits in rice and mustard, resistance to karnal bunt and rust in wheat and an extensive research network on rice biotechnology, which includes work on genomics, mapping, improvement in nutritional quality and resistance to biotic and abiotic stresses. Another multi-institutional project on tagging of quality traits in wheat by molecular markers has given some interesting leads. A number of markers linked to grain protein content, seed size, bread-making quality, etc. have been identified.

Transgenics being developed in the country both by the public and the private sector are listed in Tables 2 and 3. To further strengthen research in the area of crop biotechnology, a new institute, National Centre for Plant Genome Research (NCPGR) has been established in New Delhi with a mandate to strengthen plant biotechnology research in India.

EMPHASIS ON GENOMICS

Rice production in India holds the key to the food security of the country. After detailed consultations with experts at the National and International level, DBT and the Indian Council of Agricultural Research (ICAR) jointly took the initiative to participate in the inter-national rice genome-sequencing programme (IRGSP) in June, 2000. Two laboratories in India, one at the University of Delhi South Campus and the other at the Indian Agricultural Research Institute (IARI) in New Delhi are jointly sequencing part of chromosome II (region between 57.3 cM and 110.1 cM of the genetic map). This region carries genes for a number of biotic and abiotic stresses and genes encoding some agronomically important traits. Around 15.58 Mb of DNA has already been sequenced till December 2002. A high quality draft sequence of rice genome by IRGSP has been declared completed by the Japanese Prime Minister Junichiro Koizumi on 18 December 2002 in Tokyo. In order to take advantage of sequence information in rice, a workshop was organized in May 2002 to identify areas which should be funded for functional genomics of rice. It was decided to develop research progammes in the areas of:

(i) salinity and drought stress,
(ii) biotic stresses,
(iii) mapping of loci contributing to high yield.

The project on genome sequencing will also allow rapid progress in the characterization of allelic diversity available in *Oryza sativa* and related species and genera for resistance to diseases and pests of rice.

FUTURE DEVELOPMENTS

In the 10th Five Year Plan (2002–2007) the following areas will be given high priority:

- Rice genomics with particular emphasis on the study of genetic variability for achieving both yield increase and stabilization in the crop.
- Molecular maps in rice, wheat and grain legume crops. To date, a molecular map with reasonable density (around 1000 AFLP markers) has been developed only for mustard (*Brassica juncea*). In wheat, tagging of some interesting traits has been accomplished. The work on cereal crops will continue and mapping work will be initiated in grain legume crops. Molecular maps will be used for mapping of QTLs for abiotic stresses, plant architecture, and yield and mapping of loci conferring resistance to pests and pathogens.
- Transgenics developed in various crops through DBT funding will be tested under appropriate containment conditions and if found useful will be subjected to nutritional and toxicological tests to determine their suitability for cultivation in the farmer's field.
- Research on development of pollination control methods for the production of hybrid seed will be further strengthened. Fixation of heterosis through apomixis will be another area of research.
- Particular emphasis will be given on an improvement in the nutritional status of some of our major crops. To begin with, the three target crops will be rice, wheat and potato. High-yielding rice varieties grown in India will be either transformed with constructs that would code for high *â* carotene content or by back-crossing these varieties to 'Golden rice' transgenics already developed in Switzerland. Improving the iron content in rice grains will be another thrust area. In potato, nutritionally enhanced transgenic Indian varieties of potato carrying the *ama1* gene are already under advanced field trials. If found suitable, these will be released for cultivation.

Transgenic approaches to senescence retardation in fruits will be a major thrust area. It is important to reduce post-harvest losses in fruits and vegetables as under tropical and sub-tropical conditions senescence is much more rapid and energy costs of cold-storage are very high.

BIOSAFETY CONSIDERATIONS

Safety of transgenic material is a major concern. A major contribution of DBT has been to set up, in collaboration with other government departments, a framework for the evaluation and eventual clearance of transgenic crops for field cultivation. The release of transgenic crops is governed by the Indian Environment Protection Act (EPA) – 1986 which came into force from May 1986. The act provides a framework for the protection and improvement of environment. Later, the rules and regulations for the manufacture, use, import, export and storage of hazardous microorganisms, genetically engineered organisms or cells were notified under EPA 1986 on 5 December 1989.

A mechanism based on interaction between committees and different departments of Government of India has been set up (Figure 1). Such materials will have to meet with the approval of the following Committees: Institutional Biosafety Committee (IBSC), Review Committee on Genetic Manipulation (RCGM) and Genetic Engineering Approval Committee (GEAC).

Under the EPA 1986, the GEAC examines from the viewpoint of environmental safety and issues clearance or the release of genetically engineered organisms/ transgenic crops and products into the environment. The GEAC also grants permits to conduct experimental and large-scale field trials which are beyond the limit of 20 acres. In case of transgenic crops, applicants are required to seek clearance from the Ministry of Agriculture, which will accord the final clearance of transgenic crop varieties for agricultural activities.

Bt cotton was released for commercial cultivation in March 2002, following the biosafety procedures laid down by the Government of India. Three cotton hybrids have been granted permission for field sowing and are currently in the fields in six states namely, Maharashtra, Gujarat, Madhya Pradesh, Andhra

Pradesh, Karnataka and Tamil Nadu. Transgenic mustard hybrids are also in advanced field trials and if found suitable, may be released for general cultivation.

CONCLUSIONS

Regulatory mechanisms for the release of transgenics are in place. A Plant Variety Protection Bill (PVP) has been cleared by the Indian Parliament and will soon become operational. These developments should encourage industry to invest in the seed sector. The Government of India and the Department of Biotechnology are committed to fund research in the area of crop biotechnology. The support from the Government for funding research in crop biotechnology has increased. If tangible products come through, given the importance of agriculture to the economy, prospects of higher financial outlays for crop biotechnology are bright.

To address the general concerns of public on genetically modified food, it is of utmost importance that the regulatory system that has been established is followed strictly. The environmental effect of transgenic crops should be assessed and compared with non-transgenic crops grown with conventional practices. The field performance and drift of transgenes, either to wild relatives of the crop or to non-transgenic varieties will have to be monitored over a considerable period of time. While introducing transgenic varieties, every effort will have to be made to maintain the conventionally bred varieties for use as germplasm for future plant breeding.

To conclude, the Department of Biotechnology in collaboration with other government agencies and the public and private seed sector wants to fully exploit the opportunities provided by crop biotechnology to advance Indian agriculture towards better quality and higher overall productivity which can be sustained for many decades to come.

2

Significance of Agricultural Biotechnology

NEW TOOLS OFFER NEW OPPORTUNITIES

Human intervention for the improvement of crops, trees, livestock and fish is nothing new. For millennia, humans have bred, crossed and selected those varieties, ecotypes and breeds that were more productive, better adapted or particularly useful.

Conventional breeding practices can now be complemented by a number of new and powerful techniques. Some of these allow, for example, the propagation of plant material in glass tubes to keep it free of diseases, and the production of more sensitive and specific reagents for diagnosing diseases in plants, livestock and fish through tissue and cell culture. Others, often referred to as molecular methods, enable scientists to see the layout of the entire genome of any organism and to select plants and animals with preferred characteristics by "reading" at the molecular level, saving precious time and resources.

Modern biotechnology also includes an array of tools for introducing or deleting a particular gene or genes to produce plants, animals and micro-organisms with novel traits. This kind of genetic manipulation is called "genetic engineering" and the product is a genetically modified organism, or GMO. Both traditional and modern biotechnologies result in plants, animals and micro-organisms with combinations of genes that would not have come about without human intervention. It has to be emphasized, however, that biotechnology includes a range of techniques and products, and GMOs are but one of them.

"With the increasingly limited amount of new land available to agriculture, modern biotechnologies could complement and improve the efficiency of traditional selection and breeding techniques to enhance agricultural productivity," says Mahmoud Solh, Director of FAO's Division of Plant Production and Protection.

SO WHAT'S NEW?

A plant or an animal resistant to a particular disease can be produced through a "traditional" breeding programme, that is, through crosses with resistant relatives, selection and backcrossing again, or by the introduction of a gene that confers the resistance through genetic engineering. While the products of both approaches will be disease resistant, only the second one is a GMO. What is new is the ability of scientists to unravel the genome to look at the genes of an organism, and then make use of that information to change the organism, and even transfer genes to another organism very distant in the evolutionary scale. And that is where the controversy comes in.

"FAO recognizes that genetic engineering has the potential to help increase production and productivity in agriculture, forestry and fisheries," says FAO's Statement on Biotechnology. "It could lead to higher yields on marginal lands in countries that today cannot grow enough food to feed their people." But, it adds, FAO "is also aware of the concern about the potential risks posed by certain aspects of biotechnology. These risks fall into two basic categories: the effects on human and animal health and the environmental consequences."

These new tools offer new opportunities for solving agricultural problems where traditional techniques have failed. Genetically modified products are usually developed and used for large-scale commercial interests, and with a few exceptions, small-scale farmers have so far not benefited from the technology.

CROP BREEDING: THE GREEN REVOLUTION AND THE PRECEDING MILLENNIA

Ever since human beings began to practise settled agriculture more than 8 000 years ago, they have selected which plants to grow — first from the wild and then in cultivated fields. Early farmers chose plants that not only grew well but that seemed to have the best resistance to changing weather conditions, pests and diseases. The plant populations selected by these farmers are the basis of today's crops that feed the world. In addition to wild plants and landraces, there is a third type of agricultural plant: those bred on a research farm, either commercial or public. Plant breeders aim, through crossing and selection, to produce varieties that have desirable characteristics, such as higher yields, resistance to pests and diseases or better adaptation to their environment. The seed and planting material of these varieties are then supplied to farmers.

Breeders need access to plant genetic material with the characteristics they seek, including wild relatives and landraces-that is, to biodiversity. "The public needs to understand why biodiversity is so important," says Jose Esquinas-Alcazar, Secretary of FAO's Commission on Genetic Resources for Food and Agriculture. "It's not just about breeding more productive crops — it's about protecting and improving our food security. In biodiversity, we can find the genes that confer resistance to new threats, such as pest epidemics and climate change, as they occur."

THE GREEN REVOLUTION

The impact of scientific crop breeding became more prominent after the 1960s. Around that time, Norman Borlaug, a US scientist working in Mexico, and his colleagues used it to breed new wheat varieties that had higher yield potential and were more responsive to inputs such as fertilizer and irrigation water. Until then, attempts to increase productivity of existing local varieties had been only partially successful, as giving them too much fertilizer made them outgrow their strength, so that they toppled over.

After years of painstaking work, Professor Borlaug crossed the local wheat with Japanese dwarf varieties to produce plants that could productively utilize greater amounts of fertilizer. The resulting wheat varieties were credited with averting the mass starvation that faced the developing world in the 1960s. Those wheat varieties have been adopted and grown widely, particularly in India, Mexico and Pakistan. Professor Borlaug won the Nobel Peace Prize in 1970. Meanwhile, scientists extended the principles of modern crop breeding to other staple crops, such as rice. The Green Revolution had begun.

AN "EVERGREEN REVOLUTION"?

Beyond any doubt the Green Revolution played a critical role at that time. Recently, however, the Green Revolution has been much analysed. Some argue that it led to unsustainable use of agrochemicals and a high level of inputs. It has also been said that new plant varieties displaced landraces in the fields and thus led to a loss of biodiversity.

Nevertheless the Green Revolution was a breakthrough that contributed to food security and the fight against hunger, particularly in Asia and Mexico. With new technologies, the science of plant breeding is changing rapidly. The genetic improvement of food crops needs to continue at a pace sufficient to meet the needs of the 8.3 billion people projected in 2025, and Professor Borlaug believes that both conventional breeding and biotechnology methodologies will be needed.

The time has come for the next step. M.S. Swaminathan, an Indian scientist and winner of the 1987 World Food Prize, who was the other driving force behind the Green Revolution, has called for an "evergreen revolution". This would address a wider variety of issues and would more closely target the poorest. Dr Swaminathan believes that this effort will require more science, not less.

GOING DOWN TO THE GENES

The techniques of molecular biology, which are the basis of modern biotechnology, allow scientists to work on the individual units of heredity, the genes, understand their nature, edit their sequences-or just look at them.

In conventional breeding, the breeder needs to observe and record the performance of the different plants and animals before selecting them and then conduct several rounds of crossing and selection to get the desired varieties or breeds.

However, nowadays scientists can produce high-yielding, more nutritious crop varieties more efficiently, simply by reading the gene sequence, marking it, matching it with traits observed in the field, and by using this information, they can select thousands of individuals with desired traits in just one experiment. This process is called DNA-marker assisted selection, or MAS. It saves time.

AN EVOLVING TECHNIQUE

In crops, the efficiency of the selection process has been enhanced by the use of MAS. The International Maize and Wheat Improvement Centre (CIMMYT) in Mexico, where Norman Borlaug and his colleagues developed the dwarf wheat varieties of Green Revolution fame, are using MAS to develop resistance to a number of wheat diseases.

The International Centre for Agricultural Research in the Dry Areas (ICARDA) in Syria is using MAS to select high-quality, drought-tolerant specimens of durum wheat, which is used to make pasta, as well as cold-tolerant varieties of lentil, and blight-resistant chickpea varieties. But MAS is often not very cost-effective and so work is under way to streamline laboratory techniques to lower costs.

GENETIC MODIFICATION: REDRAWING THE MAP OF LIFE

Changing the genetic blueprint is somewhat analogous to writing computer codes. Lines of code define the behaviour of a program through sequences of 1 and 0. If you open a file containing such a code and cut and paste new sequences of 1 and 0 into it, the program will do something different.

And if you change the DNA sequence, you may change the characteristics and behaviour of the organism it defines. In practice this is done with biotechnology tools rather than a keyboard-chemicals, bacteria and even tiny golden bullets are used to insert genes into a sequence.

More conventional forms of genetic manipulation can sometimes cross organisms that wouldn't breed together in nature. But molecular biology goes further-it can introduce genes from very different and evolutionarily distant organisms. It can thus have novel applications-such as an experiment that introduced a gene from an Arctic flounder into a strawberry so that it could briefly tolerate extremely low temperatures.

REDUCING EXTERNAL INPUTS

But most GMOs are less spectacular. More typical is a successful project to make maize produce its own insecticidal protein, which would otherwise be obtained from a bacterium, *Bacillus thuringiensis.* This reduces the need to spray insecticide on the field. The same approach is being used with cotton and is beginning to produce real health benefits for agricultural workers, who are normally exposed to high levels of chemicals with this crop. Another example is soybeans modified to tolerate herbicides aimed at weeds.

Such maize, soybean and cotton crops are now being grown commercially. In fact, in 2001, the area planted with genetically modified crops was around 50 million hectares worldwide. Besides the United States, 14 countries are growing GMO crops, including Argentina, Australia, Canada, China, India and South Africa.

Those in favour of GMOs argue that the approach is not qualitatively different from other forms of genetic manipulation. For example, using controlled environments to cross two plants that couldn't meet in nature by other means also produces combinations of genes and characteristics that would otherwise not exist. Opponents of the technology counter that GMOs are different, because the techniques used to "edit" genes or combine genes of unrelated species could lead to undesirable traits due to unpredictable gene interactions and give rise to potential negative impacts on human health and the environment. Only time, careful research and a mass of data will resolve the dilemma of GMO safety.

TREES: LONG CYCLES, DEEP QUESTIONS

Forests are not the same as fields, and trees take longer to grow than crops. In forestry, genetic modification raises specific issues.

FAO has predicted that worldwide wood demand will have increased by 25 percent between 1996 and 2010. This increase won't be met from natural forests – 35 percent of the world's industrial wood comes from plantations, although they made up only 5 percent of total forest cover in 2000.

Plantation area will grow to meet demand. Globally, it increased fourfold between 1990 and 2000, and that trend is expected to continue, especially in developing countries. Where will the land come from?

"It is impossible to ignore any sound strategy for making plantations more productive," says FAO forest geneticist Pierre Sigaud. "Such strategies might include genetically modified trees. But they also include better seed selection and management practices."

And where genetic improvement is needed, he points out, there is far more untapped potential in existing forest genetic resources than in agriculture — where the crops have largely been domesticated and there are fewer wild relatives on which to draw for breeding purposes. So GMOs could have a role – but their use would be limited to plantation forestry, while, globally, the majority of forests will remain natural forests.

LONGER LIFE CYCLES MAY MEAN MORE INSTABILITY

The longer an organism grows, the greater the opportunity for variations in its environment. As a result, genetically modified trees present unique challenges. The long life of a tree means that it is more likely to encounter stresses – such as weather and pests – that could trigger unpredictable genetic responses. The slower growth of trees means that problems will take longer to show.

One risk is related to the spread of modified genes into natural populations, as tree pollen disperses widely. On the other hand, plantations often rely on introduced species that have no wild relatives growing nearby – limiting the opportunities for crossing with wild populations.

But the differences between trees and crops cut both ways. "With crops, you can adapt the environment to the plant with agronomy and agrochemicals," says Dr Sigaud. "But forests take years or decades to mature, and such practices are often uneconomic and in any case are often regulated. So a tree has to be suitable for its environment, whether it's genetically modified or not. There are few species for which genetic modification can bring significant advantages compared to traditional selection and improvement programmes.

In any case, commercial constraints limit opportunities for genetic modification of trees. Growing and harvesting a plantation takes time and money, as do trials. Because of this and concerns over public opinion, genetically modified trees so far have not been planted commercially, although there has been plenty of research – especially in high-yield plantation species.

OPPORTUNITIES AND DANGERS

Besides faster growth, qualities anticipated from genetically modified trees include: better tree form; more uniform wood quality; lower or modified lignin content; pest resistance; herbicide tolerance; sterility; and adaptation for harsh environments.

If genetically altered trees are released commercially, the first modifications are likely to be for modified lignin content, pest resistance and herbicide tolerance.

- Lignin is a heavy tissue that gives strength to plant cell walls and helps support the tree. Breaking it down during pulp and paper production requires costly, environmentally hazardous chemicals. Trees with reduced lignin could be processed more cheaply and cleanly. But they may have less resistance to pests and high winds.
- Pest resistance has already been bred into commercially released genetically modified food crops. It causes them to produce toxins that kill pests and then degrade quickly in the environment, thus causing limited direct harm to the environment. The same technique could work for trees, although there are concerns that their long life cycle could give pests more chance to develop resistance to the toxins.
- Herbicide tolerance could permit more targeted application of herbicide, thus reducing the amounts used. But this application is less important than with crops — in forestry, herbicides are generally used sparingly and only when a forest is being established.

Further ahead, there could be more environmentally friendly applications. Trees can prevent or reverse land degradation, and tree improvement, whether by genetic modification or not, could bring new opportunities of this type. Trees bred for harsh environments could help protect fragile land on desert margins, while those bred for salt tolerance could help restore the productivity of land damaged by poor irrigation practices. Also, faster-growing trees – in addition to their commercial possibilities – could help fight global warming by taking more carbon dioxide from the atmosphere and converting it into plant matter – a process called carbon sequestration.

A SAFE TECHNOLOGY?

While wood is not consumed by humans, other parts of forest trees do get into the food chain, through the consumption of pollen, honey, wild fruit and game, for example. Long-term environmental risks may be more easily overlooked, especially as forest plantations may be remote and not closely supervised. In any case, all newly introduced 'natural' or improved tree species need careful supervision, independently from genetic modification.

Making GMO trees sterile so they do not breed and release genetically modified material into the environment is an important safeguard, but it may not be completely reliable. Trees without reproductive features might reduce the diversity of birds and insects that feed on them. These questions call for caution, as should the cost and effort of investing in such technology and the time needed to get that investment back.

A further disincentive to genetic modification in forestry is the wide, largely unexplored diversity of existing forest species. Much of this can be used to make forests and forest plantations more productive, using traditional tree improvement and breeding techniques combined with sound silvicultural practices.

Commercialization of genetically modified trees could come and may make trees even more useful in a number of ways. But the technology is not likely to be adopted as quickly or as broadly as in other sectors.

AGRICULTURAL BIOTECHNOLOGY IN INDIA: ETHICS, BUSINESS AND POLITICS

Development of sustainable pest management strategies in agriculture has become necessary in view of increasing non-viability of chemical based approach. Among various approaches for the purpose, policy makers have paid far more attention to biotechnological alternatives. The first transgenic variety was approved for commercial trials in cotton.

However, some companies incorporate the Bt gene from the approved varieties in other varieties and released such hybrids to the farmers. It so happened that these hybrids though illegal and released unethically proved more remunerative to the farmers. As if this was not enough, farmers made crosses of this hybrid developed by Nav Bharat Seed Company and developed their own locally suited varieties.

The paper describes the ethical, business and political dimensions of agriculture biotechnology in India with specific reference to the experience of Bt cotton in Gujarat. The neglect of IPM, herbal pesticides and bio control methods becomes even less justified when state not only tolerates but also encourages widespread experimentation of Bt cotton without any regulation or monitoring. Implications for future policy for technological change have been outlined in the paper.

This paper deals with the case of Bt cotton in Gujarat, India, as an illustration of how public policy chickens out when large-scale violation of ethical and scientific norms takes place with positive business outcomes.

People have begun to take technological decisions into their own hand. Entrepreneurial spirit has created a unique case of farmer participatory research where on farm crosses are being made between Bt cotton varieties (including the ones which have been released without approval by the regulatory agencies) and other released varieties. New production cycles are being created by farmers who have extended the life of the crop from six months to nine months to reap advantage of continuous flowering and thus higher yield.

All this has happened in a completely unauthorized manner, with full public knowledge and despite complaints of Monsanto and MAYHCO about Nav Bharat Seed Company having 'stolen' their Bt gene. Farmers are happy, politicians do not care and regulatory agencies are satisfied that sending a few committees to enquire is all that they needed to have done.

This article presents a review of important studies in the context of the Bt adventure in Gujarat and discusses the findings obtained in a farmer survey on experience and perception of Bt cotton in Gujarat.

In a developing country with overflowing food grain stocks, a liberalizing economy set on a growth path, and techno-bureaucracy willing to pursue new technologies with a reasonable open mind, the situation seems ripe for agricultural biotechnology. The investment in this sector is picking up and the private sector has been given a pivotal role along with synergistic public investment.

Top Indian policy makers argue that biotechnology will provide food security to the hungry but do not explain as to why should hunger exist at all in India with its overflowing food stocks, if the issue was only supply! Activists also make a case for biotechnology so long as it increases productivity. The case of Bt cotton in Gujarat shows that this new technology indeed increases productivity, profits of the farmer and eliminates the need for excessive sprays of chemical pesticides.

Yet, when good economic returns follow a bad ethical practice in terms of technological change (not respecting environmental regulation and monitoring, nor intellectual property rights), do ends justify the means? The answer begs for possible alternatives to Bt cotton:

- *The agro-chemical industry and conventional agriculture*: the chemical pesticide industry makes its living so to say, by heavy inputs in cotton (reportedly 40-50 per cent of all the pesticides in India is consumed by hybrid cotton alone). It does not even educate workers and farmers about (a) safe ways of spraying (not almost a single hoarding in the entire country on the subject though there are thousands of bill boards advertising the use of the chemical pesticides), (b) disposing pesticides containers (except in small print in the publications accompanying the pesticide bottles), and (c) practicing integrated pest management –IPM-practices. Ethical violations in chemical pesticide industry are more conspicuous, have hazardous results and are more certain than other new technologies-in fact consequences of excessive use of chemical pesticides on human and environmental conditions are well documented as against the environmental consequences of biotechnology which are still being probed Yet, similar public debate does not ensue.

Integrated Pest Management (IPM): Despite having achieved excellent results through publicly funded IPM, the investment on IPM in Gujarat remained static for the last three years-around 75000 US$ per annum that is less than one percent of the total agricultural budget in Gujarat. This is very likely to be the case at the countrywide level as ascertained in informal dialogues with senior policy makers in the Ministry of Agriculture. IPM offers an alternative to biotechnological solutions in pest control and a cheaper alternative at that. But public policy support for IPM is quite weak.

Even though the adoption of Bt cotton in Gujarat has benefited farmers significantly, the public policy neglect of safer, cheaper and more affordable technological alternatives has obvious ethical implications too. The Department of Biotechnology (DBT) of the Government of India is aware of these alternatives and has made hardly any investment in making even fair comparative trials of Bt cotton with other competing technologies. Not to mention the almost total absence of investment in public education about the likely hazards of taking up such large scale un-guarded, un-monitored and un-evaluated trials by farmers of the Bt crosses with local hybrids.

Undoubtedly, the experience of Gujarat will be recalled in the history of biotechnology as one of the largest trial (with full public knowledge and without any responsible monitoring or evaluation by public agencies at similar scale) of an illegally released technology ever done by people themselves, oblivious of any environmental or other consequence.

THE CONTEXT FOR BT ADVENTURE IN GUJARAT

Historical review of the introduction of Monsanto Bt cotton and the emergence of generic products of Bt cotton through Indian companies

1990: Monsanto started negotiating the technology transfer agreement with the Government of India for its Bt cotton package.

1993: The talks failed, as agreement on financial terms of transfer was not reached.

1993: Monsanto started negotiating the technology transfer agreement with the MAHYCO for its Bt cotton package.

1995: DBT (Department of Biotechnology) approved MAHYCO to import 100 grams of cotton seeds containing Bt Cry 1 Ac gene.

1996: Approval by Central Govt. for import of the first Bt cotton variety US Cocker-312. This variety was crossed with the elite Indian varieties to produce locally adapted Bt cotton varieties with Cry1Ac gene.

1996: First trials of Bt cotton.

1997: Feild trials permitted in five states viz., AP, Karnataka, TN, Haryana and Maharastra.

1998: Field trials extended to four more states viz., MP, Rajasthan, Punjab and Gujarat.

1998: Monsanto acquired a 26 per cent stake in MAHYCO, which later became 50-50 joint venture MAHYCO-Monsanto Biotech (MMB).

1998: MMB received approval from Review Committee on Genetic Manipulation (RCGM) of DBT to conduct countrywide-field trials on 85 hectares and to produce seeds on 150 hectares.

1999: RFSTE (Research Foundation for Science, Technology and Education), New Delhi files a Public Interest petition challenging "legality" of RCGM (Review Committee of Genetic Manipulation) under DBT to approve field trials as it is not the concerned authority and also no biosafety regulations were exercised.

2000: DBT allowed Mahyco to conduct extensive field trials, including seed production at 40 sites in six states based on the "total confidential" data from small-scale trials.

2001: MMB approached GEAC (Genetic Engineering

Advisory Committee) for commercial release of Bt cotton varieties.

2001: GEAC approved field trials for another year on 100 hectares in seven states.

2001: GEAC withheld environmental clearance of large-scale cultivation of transgenic Bt cotton in June 2001. Instead it calls for fresh large-scale field trials under the direct supervision of committee set up by ICAR (Indian Council of Agricultural Research) with representatives from MOEF (Ministry of Environment and Forestry), DBT, Dept. of Agriculture and Cooperation and Ministry of Health through Advanced Varietal Trial program of All India Coordinated Cotton Improvement Project. Besides these the GEAC in the press release said "MAHYCO may like to conduct field trials on farmer's field in an area of about 100 hectares under close supervision of GEAC and Monitoring and Evaluation committee. It also advised collection of the complete evidences and data pertaining to impact of transgenic on human and animal food, spread of the cry protein resistant boll worm and impact on non-target soil microflora and other fauna.

October

2001 MMB discovers commercial cultivation of Bt cotton on over 10,000 acres (4000 hac.) in Gujarat, traced the sale of the seeds (Navbharat 151) to Hyderabad based Navbharat Seeds Pvt. Ltd. MMB also discovered that Navbharat has been in the business of selling the seeds for the previous three years and demanded punitive action against the company.

2001 Following this the GEAC ordered the Gujarat Biotechnology Coordination Committee to burn all illegal plantations.

2001 GEAC on insistence from the Gujarat Government ordered that the illegal cotton be procured at a suitable price. But the order was late as the produce was already in the market by that time.

November

2001 Case registered with Gujarat High Court against Navbharat for violating the **EPA** (Environmental Protection Act) rules.

January

2002 Secretary DBT announces that latest rounds of Bt cotton field trails were satisfactory, based on report from ICAR.

March

2002 Centre approves three years commercialization of three Bt cotton varieties – Bt MECH-12, Bt MECH-162 Bt and Bt MECH-184 with a set of conditions. It disapproved MECH 915 Bt cotton seeds.

May

2002 Rasi seeds Company Ltd. gets nod from DBT to conduct trials.

August

2002 Karnataka government bans sale of Bt cotton seed temporarily.

A BRIEF LITERATURE REVIEW OF THE CURRENT BT COTTON DEBATE

Technological change in tropical developing countries has been studied from institutional, socio-economic and cultural, policy and structural perspectives. Incentives for farmers, it has been realized, are the prime movers for technological change in most cases. These incentives or benefits may sometime be high in short term but may have low returns in the long term (as in the case of chemical intensive approach to agriculture).

From the portfolio perspective, risk and return are two obvious determinants of this change. There can be four combinations, high risk, high return; high risk, low return; low risk, high return; and low risk, low return. In the case of Bt, the risks associated with the technology are

(a) *environmental:* When the pest infestation goes down, the possibility of birds and other predators getting attracted goes down. To that extent, there is an impact on the avian biodiversity. Whether this effect is lesser or more than the pesticide sprayed cotton is a matter for further investigation. The possible effect on soil microorganisms, on other species through diffusion of Bt gene including weeds are other environmental implications.

(b) technological, i.e., resistance development of pests to the Bt gene. Such risks associated with resistance are being investigated. There does not seem to be conclusive evidence. Some argue for precaution on this ground.

(c) consumer health such as possible allergenic role of the food products (in this case, the oil of cottonseed). These risks have not been investigated adequately in the Indian context

(d) *socio-economic:* The socio-economic effect on labour by way of suspected decrease in demand, has recently been found to be offset by increased duration of the crop from six to nine months. This increased demand for labour used in picking cotton. The labour finds their productivity going up in Bt cotton compared to conventional hybrid cotton. Reportedly the number of balls infested with pests being higher in the later case makes the task of picking good quality cotton bit tedious. The labour contractors also reportedly benefit because of the significantly higher volume of cotton being picked up by the labour everyday enhancing their own margins.

(e) ethical, implying doing things with inadequate information about right or wrong outcomes on various stakeholders, imposing in the process, costs on future generations. The ethical issues are extremely serious not only because of the decisions taken with inadequate information but also because of the possible hazards in natural eco-system. The constrained choice of future generations in terms of cotton varieties without Bt gene also involve a moral dilemma. Robinson recalls that every technological revolution in the past has had major impact on the perceptions of the public about right and wrong and consequently the legitimacy of technological change. The discourse on usefulness of genetic

> engineering will have to be guided by better science but also more widespread debate. Scientists cannot complain that their confidence in the technology is being questioned by people whose understanding of science is limited. In fact, the farmers who have crossed Bt cotton variety released illegally with the local hybrids and have found economically beneficial results are paradoxically not worried about environmental implications. Most of the farmers we talked to seem to have faith in the ability of the scientists not only to monitor that but also to educate them as and when situation so warrants. The fact that scientists have delivered better hybrids in the past to overcome the resistance as well as other problems seems to assure the farmers that they would get the answers to any possible problem that transgenic cotton might create in future. The irony is that scientists themselves are not doing much systematic monitoring of the large-scale farmer participatory Bt research going on in Gujarat. For once, the scrutiny of science is not being matched with the popular faith in its capacity to deliver. The regulatory framework regarding biotechnology in India is summarized in Annexure one. It is a different matter that popular faith in this framework is not as strong as one would expect in a democratic society.

Regulatory efforts with regard to Bt cotton in India and quoted officials of the Department of Biotechnology admitting their lack of knowledge regarding the long-term safety. Scoones believes that by not engaging the critics in the science policy network, 'a sense of disengagement, distance and distrust within the regulatory process emerges.' Many agricultural and ecological scientists have felt excluded from the regulatory process. He quotes senior scientists involved with DBT about the doubt they have in the ability of DBT to enforce their own guidelines. He argues for some forms of accountability outside the science – industry – policy networks to make current discussion on biosafety regulations meaningful.

Herring, though criticizing the nature of dialogue between policy makers and the disadvantaged communities nevertheless wonders whether scarce resources are to be spent in building a more strict, widespread and bureaucratic regulatory regime or if these resources ought to go to public health, education or extension. Excessive caution, he suggests, may drain the resources required from alternative uses. He adds that insufficient caution might generate unanticipated harms.

Kinderlerer and Adcock raise the question of justice involving introducing new technologies and imply that if poor benefit, then risk might be worth taking.

In Gujarat, small and large farmers (with or without effective irrigation) seem to have voted for Bt, including the illegally marketed one, disregarding other consequences. Does it make it just? Do ends justify the means?

Gupta and Sinha have asked this question in the context of environmental governance and have wondered whether the ethics of not investing in alternative ways of solving a problem makes even more profitable outcomes from a particular approach, more legitimate and moral. Thus, if we ignore IPM and we assume that Bt cotton gives more profit than IPM approach, will Bt approach become more ethical just on that ground, regardless of overall consequences for environment.

Bharathan believes that in the case of Bt cotton, the democratization of knowledge, technology and discourse would require greater participation of the scientists from different disciplines. Venkateshwarlu had argued that despite general governmental sympathy for the farmers who violated GMO regulations, the real solutions lay in creating wider awareness about biosafety regulation rather than only making regulations more strict.

In a recent report, Jishnu referred to seed industry estimates that nearly half of the Gujarat's fields were planted with Bt cotton. She also looked at government figures about seed supplyfrom public sources, and from legal channels and the acreage under cotton to estimate the area under illegal Bt cotton.

Farmers in Gujarat have no qualms to say that the issue is not whether Navbharat Bt or not, the issue is how to get all the seed that one needs one way or other. Seldom has a technology found so widespread approval from the users without any intervention of formal extension machinery.

Undoubtedly the politicians have supported it by expressing helplessness and claiming inability to get illegal cotton uprooted by paying compensation to the farmers. While Jishnu refers to the advantage of Navbharat Bt cotton in terms of shorter duration, farmers have gone a step ahead and found that by extending the duration for about three months, they can almost double the yield with marginally extra cost. Navbharat Seed Company sources have maintained in personal communication (March 10, 2004) that they have not sold any seed for last two years.

By implication, the entire diffusion of Navbharat 151 Bt cotton seed in Gujarat during the last two years has been achieved by farmers, traders and other seed companies, which have produced seed one way or the other incorporating the Bt gene from Navbharat 151 variety.

What is even more interesting is that several postgraduate students of Gujarat Agricultural University have become entrepreneurs by developing the seed at their private farms and thus earning while learning.

Whether their values will get so shaped in this process of illegal seed production that after twenty years when some of them would be head of the department or members of regulatory bodies, they would have perhaps less difficulty in living with legal or ethical violations. They might not even ask question about ethics in such situations.

The discourse on Bt cotton is being outpaced by the fast changing ground realities. Sahai referred to the paper by Quaim and Zilberman , which created lot of controversy. It stated that the genetically modified cotton having Bt gene had shown higher yields in the experimental plots in India. Sahai argued that the failure of Bt cotton in Maharashtra and Andhra Pradesh contradicted, "the exuberant projections of two foreign scientists publishing from an American university...".

She argued that the motivations of the editorial committee of *Science,* a very reputed journal could be considered suspect. We are worried that if such be the logic, would widespread satisfaction with Bt cotton in Gujarat during last two years vindicate the decision of editorial board of *Science.*

It is a different matter that farmers have found Mahyco Mansanto Biotech (MMB) cotton yielding higher than Navbharat 151 seed but their preference still remains for the latter because of its ability to yield for longer duration, earliness in flowering and almost one fourth the seed cost (MMB seed costs about 35 USD per 450 grams whereas Navbharat costs nine to twelve dollars for the same quantity)

THE FIELD SURVEY

Survey Outline and Methodology

A sample of 363 farmers from various parts of Gujarat except Kutchh was surveyed to collect data about the experience with Bt cotton during 2001-2002, the year when Navbharat 151 seed was formally sold by the Navbharat seeds ltd. During last two years, the seed has been not sold by this company.

But the seed has been multiplied by the farmers or other private agencies as mentioned above. The sample survey has been organized on the basis of source of seed in terms of a location and producer.

The respondents were selected at random from 75 villages from 10 major cotton-growing districts of Gujarat state. The villages were selected on the basis of scatter of the technology to capture as much agro-ecological diversity as possible.

The respondents were interview by the Post Graduate students of Gujarat Agricultural University pursuing their Masters Degree in Plant Breeding and Genetics, Agril. Entomology, Agronomy and Agril. Economics. The respondents were visited at their farms and the pre designed and tested questionnaires were used to elicit their opinions.

Since the survey is based on recall data and that too with a lag of one year, there is a possibility of some loss of information. The data pertains to the year 2002. The information collected from the respondent farmers was crosschecked by asking the neighboring farmers about their views relating to the BT cotton crop of the respondent farmer. Several farmers were hesitant in giving their names since they knew that their cultivation of Bt cotton was illegal. Once assured of anonymity, they agreed to cooperate in the survey.

Field Survey Results

We provide brief information about the sample of farmers who bought seed from distant locations, local shop, other sources with in their region and those who have made it on their own. There was not much difference among these groups in regard to age composition or education.

The largest proportion of farmers in all land size groups bought seeds from distant sources signifying a distinct preference for the BT seeds even from far off places. The main difference was that small and large farmers had sizeable section which purchased seeds from shops (which had highest productivity and were thus perhaps of best quality).

The key contrast was that while majority of marginal land holders had only less than one fourth area under irrigation, compared to those who have bought from shop and from distant places having much larger area under irrigation. Among those who has procured seeds from shops and distant places, majority had irrigated holding signifying their ability ti provide better production environment to good quality seeds.

The sample characteristics are on the basis of seed producers (companies or farmers themselves). The age and education did not make much difference to seed buying behavior. However, if we look at the ratio of branded to non branded seed purchase behavior, the ratio was 3:2 among marginal land holders while it was about 2:1 in the larger land holding farmers.

If the land holding is a proxy for better economic power, those who oculd afford, did go for better seeds. What is interesting though is that even among these more opportunistic buyers, a significantly higher proportion went for illegal Navbharat seeds rather than Monsanto's legally released Bt seeds.

Poorer farmers relied much more on low quality F_2 seeds compared to the rest. A discriminant analysis among the four groups also revealed the average yield to be lower among growers of own seed or F_2 seed by about 10 to 25 per cent.

Compared to normal cotton both MMB and Navbharat flowered earlier, took lesser days to maturity, was slightly dwarf and had higher number of balls per plant. The trend was similar in the case of farmers' own seeds as well as the F_2.

The average yield was highest in the case of MMB followed by Navbharat, farmers crossed and F_2. When one looks at the source in terms of location (Table 4) the seeds obtained from shops apparently performed the best followed by the ones bought from distant sources, from other farmers, and the minimum yield was from own seed.

The experience with different pesticides was on an average better in the case of Navbharat seed than in case of MMB or other sources. When one looks at the composition of sources, location and producerwise, among all the producers, the seed of Navbharat was obtained by largest number of people from far off places indicating maximum demand and minimum availability from the shops. In terms of farmers crossed F_2, pattern was similar except that shop was not an important source of farmers crossed or F_2 seed. There was not much difference in terms of the incidence of different pests.

The survey by and large reveals that farmers growing MMB have achieved marginally higher yield but at higher costs (since as mentioned earlier, MMB seeds costed about 35 usd per 450 gms, where as Navbharat seeds bought from shop costed about USD 9-10). Whereas in case of Navbharat seeds or its derivatives, they have still managed much higher yields than the normal hybrids at much lesser cost.

From the risk perspective, what is most remarkable is the higher standard deviation (SD) in case of Bt cotton compared to the normal hybrids.

Higher risk and higher return would make the technological choice attractive for entrepreneurial farmers but for the others, it might be inhibitory. Given the fact that the survey is based on recall data and that too with a lag of one year, there is a possibility of some loss of information. We are planning to repeat this survey among the same farmers for the current year's performance. What has been demonstrated unambiguously is that in the current year a very significant number of farmers have decided to irrigate the crop even after normal season and elongate the life by about three months with very attractive returns. Normally farmers grow potato in some parts after cotton but this year they decided to continue with the cotton, particularly the Navbharat seeds and its derivatives.

POLICY IMPLICATIONS

From a simple efficiency point of view, sufficient availability of F_1 seed duly authenticated from the shops would achieve the highest profits for the farmers. However, this possibility depends on the government's ability to screen Navbharat seed for its environmental safety and other regulatory requirements and thus make its seeds available legitimately at 25 per cent of the MMB seed. Dr. Desai, MD of Navbharat Seeds in a personal interview said "when government has already tested the biosafety and environmental safety of Cry 1Ac gene, and it has concluded that Navbharat 151 contains that gene, then why should we be prevented from the commercialization of the variety, where is the risk?"

He states that they collected the germplasm from the farmers` field in Maharashtra. This seed collected from farmers' fields apparently contained the Cry 1 Ac gene. Navbharat Seeds Ltd., crossed it with Guj. Cot 10 a selection from the Bikaner Lerma, a local variety that has a wide adaptability and is very suitable for Indian conditions.

So far as the intellectual property right issues are concerned, the conflict between Monsanto and Navbharat will have to be resolved in the courts but at the face of it, there does not seem to be a case against Navbharat from the facts available so far (since India does not permit gene patents). Dr. Desai, the proprietor of Navbharat Seeds acknowledged that their company did not have the facilities for producing genetically engineered seeds.

They made crosses with Guj. Cot. 10 in the normal process in which the Bt gene got incorporated. The choice of parents was apparently much better in the case of Navbharat than was the case with MMB. Smaller companies are generally closer to the ground and thus can respond to the farmer' needs more efficiently. However, when the seed in case is Bt which is not yet protected as per the Indian laws, the case for its incorporation in other varieties in an unauthorized manner falls in the realm of morality rather than law.

The critics have argued that regulatory processes were compromised in the case of MMB to ensure its earlier release. Moral issues are involved in such compromises too. Critics have also argued that wider participation of civil society and multi disciplinary scientists has not taken place adequately. What is even more disturbing is that after the wide spread diffusion of Navbharat seed through entirely illegal channels, no systematic monitoring, data analysis and feedback system has been put in place.

What is lesser evil? Subject Navbharat Seed to as stringent environmental regulatory tests as necessary and then make it possible for farmers to obtain proper F_1 seed so that responsibility, if any, for the adverse consequences can be fixed on the company? Or to live with the situation where almost every farmer who can afford to grow Bt cotton wants to grow it and generally with Navbharat seed obtained one way or the other.

Are we trying to make virtue of a vice? Do two wrongs make one right? Because MMB is reported to have been pushed with very poor results in some states and reasonable results in other states, without adequate public debate, Navbharat Seed Company should be allowed to have the similar advantage. After all, the farmers do not seem to be complaining at all about Navbharat seeds.

We believe that markets without morals only create circulation of short term goodies which cost the society much more in the long term. Once a society learns to be comfortable with the notion of, 'ends justifying the means', the ability of the society to appreciate the merit of 'means justifying the ends' goes down. Slowly, everybody cuts the corners and opportunism undermines sustainability.

The situation regarding the Bt cotton controversy in India with particular reference to Gujarat has assumed very different dimensions because of the large-scale dispersal of illegal seeds all over the country. Regulatory agencies were aware of the illegal transportation and did very little, if anything at all.

The problem was allowed to become so widespread that no political system could have afforded to annoy such a large farming community benefiting from a technology. There are two historical analogues of this problem. In the irrigation command area projects in which a large area is supposed to be irrigated through canals drawn from a reservoir, the secondary and tertiary canals always took longer time to be constructed than the primary canal.

Farmers in the upper reaches of the primary canal got used to getting water for three seasons of paddy or other such crops whereas the tail enders did not get enough water even for dryland crops. Subsequently when the secondary and tertiary canals got built for distribution in the lower reaches, the farmers in the upper reaches created political pressure, violated the law and continued to get as much advantage as they could get.

The second example also from the irrigation sector is that of setting up lift irrigation cooperatives along the canal by lifting water from the canal. Farmers use pump sets and lift water from canal flowing on gravity principles and start irrigating their holdings through lift canal. Such a practice is illegal and affects the interests of the farmers of the lower reaches adversely who depend upon the water in main canal flowing on gravity basis only. However, in both the cases, politics prevailed over ethics.

The question in case of Bt cotton is, whether politics facilitated by markets will entirely ignore ethics irrespective of who has violated how much ethics. We contend that to simply ignore ethics in the face of opportunities offered by one single technology is not a very healthy or reasonable position when other alternatives for pest control in cotton or other crops are available such as IPM or herbal pesticides. The ethics of not giving a fair trial to low cost, farmer-innovation based pest management strategies (see Honey Bee database at sristi.org) has remained totally out of national and international discourse on Bt cotton.

The very fact that Honey Bee Network has documented thousands of innovations in this regard over last fifteen years has been ignored by those who feel very concerned about morality, justice and fairness in public discourse. Hundreds of examples are available for the last several years at http://www.sristi.org/index.php for the purpose. National Innovation Foundation (NIF) has built a huge database of grassroots innovations and traditional knowledge (www.nifindia.org) which has been ignored by the critics of Bt cotton as steadfastly as by the supporters of the Bt cotton.

Of course both may do it for different reasons. If herbal pesticides or agronomic means of pest control can reduce the costs of the farmers and thereby avoid the need for endangering environmental security, why not. But Department of Biotechnology of the Government of India must be held as much accountable for the mess that we are in, as anybody else responsible for regulating the technology.

Such is the case when several hundred farmers have committed suicide in the last few years because they could not pay back the loans taken for growing cotton with the help of chemical pesticide. We wrote to the Chief Ministers of Andhra Pradesh and Karnataka few years ago when large scale suicide deaths were reported. We sent non-chemical pest control innovations to them suggesting a fair trial of these options to help farmers.

They didn't care to respond. The biotechnology on the other hand, has received tremendously positive response from the same Chief Ministers and Central Government officials. Is it the nature of control that corporations exercise, which makes dispensation of influence so much easier for the powers that be? Or is it the genuine advantage that biotechnology offers over other means of achieving the same results in agriculture that warrants such a policy response.

In the absence of empirical trials of chemical pesticide based cotton along side the Bt cotton and the IPM cotton, it will be difficult to make any scientific inference, one way or the other. But the lack of evidence does not deter either the critics or the supporters of Bt technology to make their case. And this to us, is the core of the current tragedy not only in India but all over the world. Ethics will indeed become efficient if the nature of discourse and rules of determining the valid evidence in scientific arguments change.

Table : SOME CHARACTERISTICS OF FARMERS USING Bt SEEDS FROM DIFFERENT SOURCES

VARIABLE		DISTANT (206)	SHOP (108)	OTHERS (26)	OWN (17)
Age	UP TO 30	20 (9.71)	7 (6.48)	2 (7.69)	1(5.88)
	31-60	166 (80.58)	88 (81.48)	20 (76.92)	14(82.35)
	>61	10 (4.85)	3 (2.78)	2 (7.69)	1(5.88)
Education	Illiterate	9 (4.37)	2 (1.85)	3(11.54)	0(0.00)
	Primary	72 (34.95)	28 (25.93)	5(19.23)	3(17.65)
	High school	86 (41.75)	41 (37.96)	11(42.31)	8(47.06)
	Graduate and above	20 (9.71)	20 (18.52)	3(11.54)	4(23.53)
Land holding	Marginal farmers (<2 ac.)	32 (15.53)	8 (7.41)	8(30.77)	4(23.53)
	Small farmers (2 - 4 hac.)	99 (48.06)	49 (45.37)	6(20.08)	8(47.06)
	Large farmers (> 4 hac.)	71 (34.47)	42 (38.89)	10(38.46)	5(29.41)
Per cent irrigated holding	< 25 %	18 (8.74)	4 (3.70)	1 (3.85)	1 (5.88)
	25- 50 %	42 (20.39)	12 (11.11)	2 (7.69)	2 (11.77)
	51- 75 %	47 (22.82)	16 (14.82)	6 (23.08)	4 (23.53)
	> 76 %	95 (46.12)	67 (62.04)	15 (57.69)	10 (58.82)
	Missing	4	9	2	0

* per cent values are worked out from total respondents as mentioned on top of the table

Table : 2 PRELIMINARY CHARACTERISTICS OF FARMERS GROWING/USING SEEDS OF DIFFERENT VARIETIES OF BT COTTON

SI	VARIABLE	Classes	Navbharat (133)	Monsanto (106)	Farmer crossed (77)	F2 (41)
1	Age	<30	10 (7.52)	11 (10.38)	7 (9.09)	4 (9.76)
		30-60	108 (81.2)	82 (77.36)	65 (84.42)	33 (80.49)
		>61	8 (6.02)	2 (1.84)	3 (3.90)	1 (2.44)
2	Education	Illiterate	7 (5.26)	3 (2.83)	2 (2.60)	2 (4.88)
		Primary	44 (33.08)	26 (26.53)	28 (36.36)	10,(24.39)
		High school	49 (36.84)	49 (46.23)	32 (41.56)	16 (39.07)
		Graduate and above	17 (17.78)	12 (11.23)	11 (19.29)	7 (17.39)
3.	Land holding	Marginal (<2 hac.)	21 (15.79)	9 (8.49)	12 (15.58)	10 (24.39)
		Small farmer (2 - 4 hac.)	60 (45.11)	39 (36.79)	33 (42.86)	20 (48.78)
		Large farmer (> 4 hac.)	47 (35.34)	42 (29.62)	30 (38.96)	9 (21.95)
4.	Per cent irrigated holding	< 25 %	10 (7.52)	5 (4.72)	6 (7.79)	3 (7.32)
		25- 50 %	23 (17.29)	16 (15.09)	12 (15.84)	7 (17.07)
		51- 75 %	26 (19.55)	23 (21.70)	18 (23.38)	6 (14.63)
		> 76 %	69 (51.87)	56 (52.83)	39 (50.65)	23 (56.10)
		Missing	5	06	2	02

* Per cent values are worked out from total respondents as mentioned on top in the table

Table : CHARACTERWISE PERFORMANCE OF VARIOUS BT CROP VARIETIES IN COMPARISON TO NORMAL COTTON

SI	Character	Navbharat F_1 (133)				Monsanto F_1 (106)			
		BT		Normal		BT		Normal	
		Mean	SD	Mean	SD	Mean	SD	Mean	SD
1	Days to 50 per cent flowering	76.96	11.52	95.39	14.72	75.95	18.79	88.92	23.53
2	Days to maturity	151.52	24.71	181.87	32.15	158.48	24.96	178.51	28.04
3	Plant height (feet)	5.08	0.99	5.85	0.88	5.44	1.01	5.78	0.96
4	Number of balls per plant	95.35	23.64	66.65	20.15	89.80	23.38	70.89	16.63
5	Flower shedding/plant	59.73	14.32	69.88	17.77	60.52	13.31	62.75	15.30
6	Yield (Kg/acre)	1229.51	520.90	800.73	304.31	1327.83	635.97	936.63	376.79

SI	Character	Farmer crossed (77)				F_2 (41)			
		BT		Normal		BT		Normal	
		Mean	SD	Mean	SD	Mean	SD	Mean	SD
1	Days to 50 per cent flowering	75.25	12.94	95.32	12.31	76.94	1057	91.77	19.09
2	Days to maturity	155.52	22.63	188.01	22.63	157.44	25.08	187.65	33.06
3	Plant height (feet)	4.86	0.95	6.23	0.99	4.79	1.07	5.74	0.79
4	Number of balls per plant	98.62	26.93	66.32	16.94	91.67	18.50	71.32	15.34
5	Flower shedding/plant	59.33	17.43	70.46	16.28	59.62	13.70	66.76	17.33
6	Yield (Kg/acre)	1198.83	400.32	775.87	219.70	1169.27	401.77	868.00	246.40

Table : CHARACTERWISE PERFORMANCE OF VARIOUS BT CROP VARIETIES FROM DIFFERENT SOURCES IN COMPARISON TO NORMAL COTTON

SI	Character	Other (26)				Shops (108)			
		BT		Normal		BT		Normal	
		Mean	SD	Mean	SD	Mean	SD	Mean	SD
1	Days to 50 per cent flowering	78.37	17.04	78.40	9.97	72.00	20.61	85.49	27.59
2	Days to maturity	153.46	26.67	188.25	35.11	155.71	28.99	168.20	25.68
3	Plant height (feet)	4.88	0.98	6.18	0.79	5.62	0.87	5.52	0.89
4	Number of flowers/plant	258.46	95.07	288.12	93.84	206.86	81.62	200.16	85.95
5	Number of balls per plant	93.84	25.23	64.79	17.47	92.17	23.44	71.01	11.15
6	Flower shedding/plant	62.42	11.74	72.37	17.04	55.66	13.45	56.53	13.45
7	Yield (Kg/acre)	1117.69	337.92	855.42	290.28	1393.38	562.91	953.98	349.51

SI	Character	Own (17)				Distant (206)			
		BT		Normal		BT		Normal	
		Mean	SD	Mean	SD	Mean	SD	Mean	SD
1	Days to 50 per cent flowering	74.68	13.35	94.66	13.94	78.03	10.50	95.76	11.56
2	Days to maturity	155.29	26.36	176.56	30.64	155.04	21.67	189.86	29.77
3	Plant height (feet)	4.97	1.19	5.75	0.70	4.9	1.00	6.06	0.94
4	Number of flowers/plant	262.05	69.95	254.06	72.55	270.38	87.52	275.27	90.14
5	Number of balls per plant	85.58	20.60	71.87	25.29	95.61	24.25	67.27	19.87
6	Flower shedding/plant	62.75	15.72	66.46	16.90	61.38	15.14	72.47	15.97
7	Yield (Kg/acre)	930.59	283.12	780.31	304.33	1209.42	519.36	789.42	281.77

REGULATORY REFORMS IN BIOTECHNOLOGY

These rules are applicable for manufacture, use, import, export and storage of hazardous microorganisms and genetically engineered organisms or cells and also correspondingly to any substances and products and food stuffs, of which such cells, organisms or tissues hereof form part.

Competent Authorities and their Functions

Recombinant DNA Advisory Committee: responsible to review the developments that take place in the field of biotechnology at national and international levels and recommend safety regulations for India.

Review Committee on Genetic Manipulation: to monitor safety related aspects of ongoing research projects and activities involving genetic engineered organisms/hazardous microorganisms. It also has to lay down the procedures restricting or prohibiting sale importation and use of genetically engineered organisms.

Institutional Biosafety Committee: will look at the experiments for the purpose of education outside the laboratory areas.

Genetic Engineering Approval Committee: responsible for approval of large-scale use, release, production and experimental field trials of hazardous microorganisms and recombinants in research and industrial production from the environmental angle.

State Biotechnology Co-ordination Committee: it is functional at state level wherever necessary. It has the powers to inspect, investigate and take punitive action in case of violations of statutory provisions.

District Level Committee: to monitor the safety regulations in installations engaged in the use of genetically modified organisms/hazardous microorganisms and its applications in the environment. Various Provisions of the guidelines:

- No person can procure and use or sell and hazardous microorganisms of genetically engineered organisms/ substances or cells except with the approval of the GEAC.
- Use of the material shall only be allowed in laboratories or inside laboratory area notified by the MoEF under EP Act, 1986.

- GEAC shall give the occupier the directions or take measures concerning the discharge of microorganisms/ genetically engineered organisms/substances or cells
- Any persons operating must obtain the license issued by the GEAC.
- GEAC may in special cases give approval of deliberate release.
- Any person applying for the approval shall submit information and make examinations or cause examinations to be made to eradicate the case, including examinations and on-site experimentation plan according to specific directions and at specific laboratories.
- Approvals by the GEAC shall be for specific period not exceeding four years at the first instance renewable for two years at a time.
- The approval can be revoked in case
    i. of any new information as to the harmful effects of the approved organisms.
    ii. if the approved organisms cause damage of environment, nature or health.
    iii. non compliance with conditions stipulated by GEAC.

The supervision will be carried out through GEAC through SBCC and State Pollution Control Board. In case of immediate intervention by the SBCC or DLC to prevent the damage to nature, environment and health they may take the action without issuing the order on the expense to be incurred by the person responsible for the damage. It is the responsibility of the DLC to prepare an off-site emergency plan detailing how emergencies relating to a possible major accident at a site will be dealt with and in preparing the plan.

The GEAC may fix fees to cover, in whole or in part, the expenses incurred by the authorities in connection with approvals, examinations, supervisions and control.

Any appeal against the decision of GEAC may be made in 30 days from the date decision was communicated.

3

Assessing Stakeholder Attitudes to Agricultural Biotechnology in Developing Countries

The current global system governing biotechnology has largely been shaped by Western hopes and fears and more research is needed into how developing countries view this new technology. This article compares two surveys that took place in the Philippines and Mexico and assesses some of the findings. Such results may be useful in developing integrated global risk and benefit management strategies.

In industrialized countries, psychologists, sociologists and political scientists have been researching public perception and the political dimensions of the agricultural biotechnology debate. However, few surveys have focused on how agricultural biotechnology is discussed and perceived in developing countries. This lack of research is one of the reasons why certain stereotypes have emerged in the North about how agricultural biotechnology is seen in the South.

Such preconceptions make it possible for extreme proponents and opponents of agricultural biotechnology in industrialized countries to select representatives from developing countries whose views and interests coincide with their own, and encourage them to speak in the name of all the local stakeholders in their country. As a result there has often been an over-simplification of the way the risks and benefits of agricultural biotechnology are seen in the South.

STAKEHOLDER SURVEYS

Between 1997 and 2001, three public attitude studies on perceptions of agricultural biotechnology in developing countries were planned and carried out in the Philippines, Mexico and South Africa. This article deals with the methodology used in these studies and discusses initial results from the Mexico and Philippines surveys. Material collected in South Africa is still being analyzed. The first study focused on the Philippines (see also *Monitor No. 38*). It was initiated by the *Swiss Federal Institute of Technology Zurich* (ETH) as part of a larger research project investigating public acceptance of genetically engineered food in industrialized and developing countries. The Philippines was chosen because it had been at the centre of a controversy surrounding *Bacillus thuringiensis* (Bt) rice, a pest resistant variety developed by ETH and the *International Rice Research Institute* (IRRI). Subsequently the Swiss National Science Foundation (SNSF) approved funds for additional case studies with similar objectives to be carried out in Mexico and South Africa. These countries were selected because they had substantial local biotechnology research capacity and were the scene of considerable controversy and debate about the role of transgenic crops in local agriculture. They also each had enough regional political and economic power to influence public attitudes in neighbouring countries.

ASSESSING PERCEPTIONS

An individual's perceptions of the risks and benefits of a new technology is determined by personally selected sources of information, values, interests, and individual experiences. However, where agricultural biotechnology is concerned, most people cannot rely on their own experience. Instead, they depend on information received from other sources such as rumours, the experiences of people working in the field, and statements issued by industry, government, public interest groups, academia and the media. An individual's personal worldview strongly determines how information is assessed for reliability.

In each of the three case studies, a stakeholder-based approach was used to investigate public perception. The stakeholders selected were political actors who were generally recognized as playing a significant role in national agricultural biotechnology debates. It was assumed that such people would be well informed and would, therefore, have considerable influence on public opinion. This approach made it possible to conduct a survey on public attitudes in spite of a low level of public awareness of the subject. It also made it possible to analyse the way perceptions, interests and scientific knowledge shape political decisions and provided a basis for predicting how the debate might develop over time. Although it can be assumed that political actors have a significant influence in countries where little is known about agricultural biotechnology, it is important to realise that sometimes they may be ill-informed themselves. In our surveys, contradictory answers and the number of 'don't know' replies reflected the extent to which political actors were informed about the risks and benefits of agricultural biotechnology.

SELECTING POLITICAL ACTORS

Political actors were selected with the help of key informants familiar with both the debate and its stakeholders. Moreover, member lists of national committees concerned with agricultural biotechnology and media coverage of the debate were screened in search of additional important stakeholder representatives. Most of those eventually selected to take part in the surveys were either from *non-governmental organizations* (NGOs) concerned with rural development and environmental issues and farmers' organizations, or government departments responsible for health, the environment, trade, agriculture, and science and technology. Other respondents included agronomists, biotechnologists, sociologists, and environmental scientists from local universities, representatives from agribusiness, organic agriculture and the food industry, the mass media and international organizations.

THE QUESTIONNAIRE

Stakeholder representatives were asked to fill in a four part, semi-standardized questionnaire. They were first asked about their perceptions of the problems facing agriculture and the potential of *genetic engineering* (GE) to solve them. Secondly, they were required to react to positive and negative statements on the risks and benefits of agricultural biotechnology. Then an attempt was made to assess how much trust they had in existing institutions and in new approaches to risk reduction and legislation.

The final section of the questionnaire included a policy network table of about 70 stakeholder organizations. Respondents were asked to assess each other on attitudes to agricultural biotechnology and their influence on public opinion, political decisionmaking and the biotechnology debate. They were also asked whether they strongly associated certain organizations with particular individuals and whether they cooperated with these organizations on matters such as information exchange and funding.

Local partners in all three countries were closely involved in drafting the questionnaires and subsequent questions clearly reflected the character of the regional debate. In the Philippines, for example, the commercial cultivation and retailing of either locally developed or imported transgenic products had been prohibited and there was a heated debate over the risks and benefits of Bt rice. The Mexican survey, however, was drafted against a background of field tests, pilot projects with both locally developed and imported *genetically modified* (GM) crops, and the discussion on introducing specific and mandatory labelling legislation to deal with imports of genetically modified Bt corn from the United States.

REACTIONS AND PARTICIPATION

A survey on public perceptions of agricultural biotechnology can easily become a political event in itself. In all three countries many political actors were suspicious of such research. Government officials were concerned about its political impact, anti-GE groups suspected hidden agendas, and scientists in private and public agricultural biotechnology research were not convinced of its usefulness. There was also some nationalist sentiment in the criticisms of those who objected to foreign research institutes initiating studies of local perceptions.

In spite of these objections and the length and complexity of the questionnaire, participation rates were surprisingly high: 65 political actors in the Philippines, 51 in Mexico and 48 in South Africa. Six months were allocated to the Philippines survey which may account for a higher participation rate there than in South

Africa and Mexico where only one month was allocated because of funding constraints.

On average, 60 to 70 per cent of questionnaires were returned in each country. Our successful collaboration with local research institutes, considered trustworthy by both sides and which were able to allay nationalist objections, contributed significantly to the high participation rate. It also emerged that stakeholder representatives were motivated to take part because they were interested to find out how other stakeholders had assessed them and whether their perceptions matched average perceptions. By persistently following up questionnaires we were able to get even reluctant representatives to participate. It is significant that those who refused to take part did so not because they were suspicious of the project but because the questionnaire took so long to complete.

DATA ANALYSIS

Data analysis consisted of a descriptive analysis of prevailing perceptions, a cluster analysis presenting an evaluation of different perception patterns, and a visual representation of the principal component analysis, which portrayed the perceptions of every single stakeholder on a two-dimensional scale. The policy network table data was subject to a simplified policy network analysis that revealed the influential stakeholders in the debate and the different types of cooperation among them. After the results had been collated, a workshop was held to present the findings, discuss the implications of the results and to evaluate the appropriateness of the survey.

INITIAL FINDINGS

Comparison of results to the questions put to respondents in the Philippines and Mexico exposed some interesting differences between the two countries. Most respondents in Mexico and the Philippines consider biotechnology to be 'just a new tool' with the potential to solve problems that currently cannot be addressed with conventional methods and to eventually contribute to future food security. While respondents in both countries disagree with the statement that GM foods pose a health risk, they expect GM crops to have a negative impact on biological diversity and are concerned that the implementation of biosafety guidelines will not be enforced properly.

AGRONOMIC PROBLEMS

Respondents in Mexico stressed the usefulness of biotechnology in combating agronomic problems such as drought, pests and plant disease that hinder agricultural productivity. Unlike respondents in the Philippines, they were more optimistic about the sustainability of GM crops and more sceptical of organic farming being a better alternative for resource-poor farmers. The efforts of farmers and NGOs in the Philippines to develop organic farming methods to ensure sustainable food security may be an important reason for the great expectations regarding organic farming.

INSTITUTIONAL FACTORS

In both Mexico and the Philippines there was concern about the corporate control of agricultural biotechnology. Personal interviews in the Philippines and attitudes towards statements regarding *Intellectual Property Rights* (IRPs) in the Mexican survey indicated that some stakeholders in these countries would prefer to create their own domestic research capacities and design varieties that were more appropriate to addressing problems in domestic agriculture rather than simply importing GM crops from industrialized countries.

Both Mexico and the Philippines are home to large international agricultural research centres, such as the *International Maize and Wheat Improvement Center* (CIMMYT) in Mexico and the *International Rice Research Institute* (IRRI) in the Philippines, which are also exploring the possibilities of GE for the genetic improvement of food crops.

As public perception and acceptance of GM crops is closely related to trust, an interesting outcome of the survey was that stakeholders in Mexico tend to consider CIMMYT a trustworthy but not very influential stakeholder in the national debate whereas this was not the case with IRRI in the Philippines. IRRI is considered to be a central actor in the Philippines and its presence in the country is considered more controversial. This difference in the perception of foreign institutions has also influenced the public debates in the countries themselves.

BIOSAFETY AND RISK

Both countries were concerned about their countries' rich biological diversity. Mexican stakeholders were particularly worried about the potential of GM maize to outcross with local indigenous maize varieties. Filipino respondents expressed concern about the effects of pest resistant rice on non-target organisms. An analysis of the Mexican survey indicated that political stakeholders seemed to be informed about the discussion on the potential health and environmental risks of agricultural biotechnology but only a few were familiar with new approaches to reduce these risks.

Respondents in both countries had doubts that their national biosafety guidelines can be implemented effectively. In Mexico, respondents considered existing biosafety regulations to be inadequate whereas in the Philippines stakeholders regarded regulatory provisions to be sufficient.

This difference may be explained by the fact that in 1991 the Philippines had already designed stringent biosafety guidelines for *genetically modified organisms* (GMOs). Mexico's regulatory biosafety framework mainly follows the USA though a recent initiative for a law on biosafety is currently under consideration, which includes the precautionary principle stated in the *Cartagena Protocol on Biosafety* (CPB).

STAKEHOLDER PERCEPTION

A cluster analysis of the data collected reflected the perceptions of various stakeholder groups. In the Philippines, for example, three clear patterns of perception emerged.

NGOs and other public interest groups appeared to be largely opposed to GE and did not think it had any potential to solve agricultural problems. A second group made up of scientists from private and public research institutes were mainly positive towards GE and saw its potential in targeting problems caused by pests, viruses and exposure to stress. In the third cluster were government officials and politicians. They saw biotechnology as having considerable potential to solve many different agricultural problems but had an ambiguous attitude toward the potential risks and benefits of GE.

In the Mexican survey, the first cluster contained not only NGOs and other public interest groups but also representatives of government agencies (concerned with biosafety and environmental issues) and some academics. This group tended to have a negative attitude towards biotechnology.

The second and largest group included scientists, government officials, politicians, NGO representatives and members of the business community and representatives of international organizations. These individuals tended to favour agricultural biotechnology and most believed in the potential of GE to solve agronomic and post-harvest problems. The third group was more strongly in favour of agricultural biotechnology and consisted of respondents from all institutional groups.

POLARIZATION

Most respondents included in the Mexican survey were scored moderate on a scale of perceptions and, generally speaking, positions adopted were less hard than in the Philippines. Filipinos were more frequently found on the outer reaches of the perception scale indicating a stronger degree of polarization. In both countries, there seems to be a significant perception gap between government departments.

Agencies affiliated to the Department of Environment tend to be more critical of agricultural biotechnology than those dealing with trade, agriculture and science and technology. Interestingly, it would appear that the national debate in the Philippines was already polarized before concerns arose in Europe in the late 1990s.

MEDIA AND INFORMATION

The policy network analysis indicated that Filipinos felt NGOs and the mass media exerted a strong influence on public opinion. Analysis of network cooperation revealed that NGOs in the Philippines play a crucial role in the acquisition and dissemination of information. Information sourced from national universities is sent to the legislature and mass media in the form of position papers. Civil society groups use public pressure created by media campaigns to influence political decisions. This was not the case to the same extent in Mexico where public opinion, although influenced by a wide spectrum of stakeholders from different institutional settings, seemed to carry less weight. Mexican NGOs and the mass media were, however, seen as being important stakeholders in the political decision-making process.

ACADEMIA

In both countries, academia emerged as the central stakeholder in the agricultural biotechnology debate. In Mexico and the Philippines academic institutions are the most important distributors of information. They are significant players in the public debate on agricultural biotechnology and enjoy public confidence, playing a crucial role in facilitating constructive dialogues between the more antagonistic stakeholders.

Although a majority of respondents from academia have a favourable attitude towards agricultural biotechnology, they can be found in all three perception clusters. The reason why some academics have a critical attitude towards agricultural biotechnology may be that they see GE as an imported Western technology and not as the product of domestic research and development. This is particularly so in the Philippines where many local scientists are more involved with international research centres and multinational corporations than with national academia.

In addition, many academics from disciplines such as social science and ecology, work closely with Filipino NGOs in the research and development of alternative rice breeding and pest management programmes. The NGO research approach is considered to be home-grown and appeals to nationalist feelings, a factor that often plays a major role in the Filipino debate.

FEEDBACK FROM THE WORKSHOPS

Workshops were held with respondents to evaluate the survey process and its results. A major criticism raised in both Mexico and the Philippines was that the perception of the farmers who ultimately have to choose whether or not to adopt transgenic crops had not been taken into account. This is a justifiable objection, but it was argued that these surveys were not intended to investigate the potential adoption rates of transgenic crops. The surveys sought, rather, to reach farmers through those who claimed to represent their interests in the national public debate. Whether these actors adequately represent farmers' perceptions and interests is a different question.

Stakeholder analysis indicated the political influence of the different farmer organizations and how much access they had to the political decision-making processes. It often turns out that resource-poor farmers are badly organized and under-represented in the national public debates. Many big national farmer organizations and NGOs claim to speak for marginal farmers but do not necessarily represent their interests well.

Workshop participants also stressed the importance of taking into account the opinions of all consumers and producers on transgenic food. However, the insistence that each opinion is equal in reason and value makes it impossible to find a common long-term policy that will minimize the risks and maximize the benefits of agricultural biotechnology by striking a compromise between different interests and perceptions. The failure to find a middle ground and design a pragmatic strategy may increase the real risks related to biotechnology, because political paralysis as an answer to uncertainty only creates more uncertainty.

In Mexico several respondents criticised the selection of political actors. The survey was conducted at a time when the long ruling *Partido Revolucionario Institucional* (PRI) had just lost the elections for the first time in 71 years and the new administration had replaced many of the political actors formerly involved in the biotechnology debate.

It is clear that there are difficulties in addressing the dynamic aspects of a public debate through surveys. However, interpreting the results of these surveys in the context of a particular political system and its cultural and historical background, may help to explain future political events and changes in perception and political influence of old and new political actors.

CONCLUSION

The aim of the overall research project was to understand the knowledge, perceptions and interests-that shape debates on agricultural biotechnology in developing countries. The two surveys conducted in the Philippines and Mexico indicate that expectations and concerns about agricultural biotechnology may differ from those prevailing in developed countries. Moreover, a comparison of stakeholder perception in Mexico and the Philippines shows that the political, cultural and historical backgrounds significantly influence the issues that are discussed at the national level. The conclusions indicate that a global system of governance of biotechnology has to take into account perceptions specific to individual developing countries, not only to manage risk, but also to ensure the fair and equitable distribution of benefits.

4

The Ecological Impacts of Agricultural Biotechnology

Transgenic crops (GMCs: genetically modified crops), main products of agricultural biotechnology, are increasingly becoming a dominant feature of the agricultural landscapes of the USA and other countries such as China, Argentina, Mexico and Canada.

- Worldwide, the areas planted to transgenic crops jumped more than twenty-fold in the past six years, from 3 million hectares in 1996 to nearly 44.2 million hectares in 2000. [10]
- In the USA, Argentina and Canada, over half of the average for major crops such as soybean, corn and canola are planted in transgenic varieties.
- Herbicide resistant crops (HRCs) and insect resistant crops (Bt crops) accounted respectively for 59 and 15 percent of the total global area of all transgenic crops in 2000.

Big business claims GMCs will reduce the use of chemical pesticides.

Transnational corporations (TNCs) such as Monsanto, DuPont, and Novartis, the main proponents of biotechnology, argue that carefully planned introduction of these crops should reduce or even eliminate the enormous crop losses due to weeds, insect pests, and pathogens. In fact, they argue that the use of such crops will have added beneficial effects on the environment by significantly reducing the use of agrochemicals. [13] However, ecological theory predicts that as long as transgenic crops follow closely the pesticide paradigm prevalent in modern agriculture, such biotechnological products will do nothing but reinforce the pesticide treadmill in agroecosystems, thus legitimizing the concerns that many environmentalists and some scientists have expressed regarding the possible environmental risks of genetically engineered organisms. In fact, there are several widely accepted environmental drawbacks associated with the rapid deployment and widespread commercialization of such crops in large monocultures, including:

Pests show rapid evolution in resisting the pesticide properties of GMCs.

Toxic buildup in GMCs harms useful insects.

- the spread of transgenes to related weeds or conspecifics via crop-weed hybridization
- reduction of the fitness of non-target organisms through the acquisition of transgenic traits via hybridization
- the rapid evolution of resistance of insect pests such as Lepidoptera to Bt
- accumulation of the insecticidal Bt toxin, which remains active in the soil after the crop is ploughed under and binds tightly to clays and humic acids;
- disruption of natural control of insect pests through intertrophic-level effects of the Bt toxin on predators
- unanticipated effects on non-target herbivorous insects (i.e., monarch butterflies) through deposition of transgenic pollen on foliage of surrounding wild vegetation
- vector-mediated horizontal gene transfer and recombination to create new pathogenic organisms

Monoculture, or farming only one crop, can lead to economic disaster and hunger.

This chapter will focus on the known effects of the two dominant types of GMCs: herbicide resistant crops (HRCs) and insect resistant crops (Bt).

BIOTECHNOLOGY, AGRODIVERSITY AND FARMERS' OPTIONS

The spread of transgenic crops threatens crop diversity by promoting monocultures which leads to environmental simplification and genetic erosion. History has repeatedly shown that uniformity characterizing agricultural areas sown to a smaller number of varieties is a source of increased risk for farmers, as the genetically homogeneous fields may be more vulnerable to disease and pest attack. Farmers have many choices, other than biotechnology, that work.

Several people think that HRCs and Bt crops have been a poor choice of traits to feature the technology, given predicted environmental problems and the issue of resistance evolution. In fact, there is enough evidence to suggest that both these types of crops are not really needed to address the problems they were designed to solve. On the contrary, they tend to reduce the pest management options available to farmers. There are many alternative approaches, (e.g., rotations, polycultures, cover crops, biological control, etc.) that farmers can use to effectively regulate the insect and weed populations that are being targeted by the biotechnology industry. To the extent that transgenic crops further entrench the current monocultural system, they impede farmers from using a plethora of alternative methods.

Are we altering the genetic structure of living things in the name of utility and profit?

ECOLOGICAL EFFECTS OF HRCS

Gene Flow

Just as it occurs between traditionally improved crops and wild relatives, pollen mediates gene flow between GMCs and wild relatives or conspecifics despite all possible efforts to reduce it. Little is known about the long-term persistence of crop genes in wild populations or about the impact of fitness-related crop genes on the population dynamics of weedy relatives. The main concern with transgenes that confer significant biological advantages is that they may transform wild/weed plants into new or worse weeds.

Hybridization of HRCs with populations of free living relatives will make these plants increasingly difficult to control, especially if they are already recognized as agricultural weeds and if they acquire resistance to widely used herbicides. For example:

- Transgenic resistance to glufosinate can be passed on from Brassica napus to populations of weedy Brassica napa, and persist under natural conditions.
- In Europe there is a major concern about the possibility of pollen transfer of herbicide tolerant genes from *Brassica* oilseeds to *Brassica nigra* and *Sinapis arvensis*.

The majority of GMCs have been engineered to repel unwanted plants or weeds.

Economic and Agronomic Implications

World-wide in 2000, transgenic herbicide resistant crops were planted on 74% of the 44.2 million hectares devoted to transgenic crops.[10] In North America, transgenic glufosinate resistant cultivars of canola and corn, and transgenic glyphosate resistant cultivars of soybean, corn, cotton, and canola are now commercially available. Bromoxynil resistant transgenic cotton has also been developed. The so-called Round-up ready soybeans are the most prevalent GMCs.

Transgenic herbicide resistance in crop plants simplifies chemically based weed management because it typically involves compounds that are active on a very broad spectrum of weed species. Post-emergence application timing for these materials fits well with reduced or zero-tillage production methods, which can conserve soil and reduce fuel and tillage costs.

Editor's Note: Sales of biotechnology products are reaching $60 billion/year.

However, HRCs also have significant problems.

- Reliance on HRCs perpetuates the weed resistance problems and species shifts that are common to conventional herbicide based approaches.
- Herbicide resistance becomes more of a problem as the number of herbicide modes of action to which weeds are exposed becomes fewer and fewer, a trend that HRCs may exacerbate due to market forces.
- Given industry pressures to increase herbicide sales, acreage treated with broad-spectrum herbicides will expand, exacerbating the resistance problem. For example,

it has been projected that the acreage treated with glyphosate will increase to nearly 150 million acres. Although glyphosate is considered less prone to weed resistance, the increased use of the herbicide will result in weed resistance, even if more slowly, as it has been already documented with Australian populations of annual ryegrass, quackgrass, birdsfoot trefoil and *Cirsium arvense.*[7]

- Perhaps the greatest problem of using HRCs to solve weed problems is that they steer efforts away from crop diversification and help to maintain cropping systems dominated by one or two annual species. Crop diversification can
  - reduce the need for herbicides
  - improve soil and water quality
  - minimize requirements for synthetic nitrogen fertilizer
  - regulate insect pest and pathogen populations
  - increase crop yields and reduce yield variance.

Thus, to the extent that transgenic HRCs inhibit the adoption of diversified cropping systems that include rotational crops, cover crops and green manure, they hinder the development of sustainable agriculture.

The number of crops engineered for insect resistance is on the rise.

ECOLOGICAL RISKS OF BT CROPS

Based on the fact that more than 500 species of pests have already evolved resistance to conventional insecticides, pests can also evolve resistance to Bt toxins present in transgenic crops. No one questions if Bt resistance will develop, the question is now how fast it will develop. Susceptibility to Bt toxins can therefore be viewed as a natural resource that could be quickly depleted by inappropriate use of Bt crops. However, cautiously restricted use of these crops should substantially delay the evolution of resistance. But is cautious use of Bt crops possible given commercial pressures that have resulted in a rapid rollout of Bt crops reaching 8.2 million hectares worldwide in 2000? Organic farmers can produce a significant yield without insecticides.

The refuge strategy of setting aside 20-30% of land to non-Bt crops to delay resistance is very difficult to implement regionally. Data from the Midwest shows that Bt corn saves on some insecticide use and yields are 2.4 Bu/acre higher than conventional corn but only under high European corn borer infestations (USDA 1999).

On the other hand organic corn growers use no insecticides and obtain yields (4.8-9 t/ha) similar or slightly higher than conventional farmers (5.0-7.1 t/ha). GMCs may have unintended victims, such as the monarch butterfly or lacewing.

BT Crops and Beneficial Insects

Bacillus thuringiensis proteins are becoming ubiquitous, highly bioactive substances in agroecosystems. Most non-target herbivores colonizing Bt crops in the field ingest plant tissue containing Bt protein which they can pass on to their natural enemies in a more or less processed form.

Polyphagous natural enemies (polyphagous: subsisting on many kinds of foods) that move between crop cultures are found to frequently encounter Bt-containing non-target herbivorous prey in more than one crop. This is a major ecological concern given previous studies that documented that Cry1 Ab adversely affected the predaceous lacewing *Chrysoperla carnea* reared on Bt corn-fed prey larvae.

These findings are problematic for small farmers in developing countries who rely on insect pest control, which involves a complexity of predators in their mixed cropping systems. Research shows that natural enemies can be affected directly through inter-trophic level effects of the toxin present in Bt crops.

This raises serious concerns about the potential disruption of natural pest control, as polyphagous predators will encounter Bt-containing, non-target prey that move within and between crop cultivars throughout the crop season.

Disrupted biocontrol mechanisms will likely result in increased crop losses due to pests or to the increased use of pesticides by farmers with consequent health and environmental hazards. Toxins from GMCs remain active in the soil, decreasing soil fertility.

Effects on the Soil Ecosystem

The possibilities for soil biota to be exposed to transgenic products are very high. The little research conducted in this area has already demonstrated:

- There is long term persistence of insecticidal products (Bt and proteinase inhibitors) in soil.
- The insecticidal toxin produced by Bacillus thuringiensis subsp. kurskatki remain active in the soil, where it binds rapidly and tightly to clays and humic acids.
- The bound toxin retains its insecticidal properties and is protected against microbial degradation by being bound to soil particles, persisting in various soils for at least 234 days.
- The presence of the toxin in exudates from Bt corn and verified that it was active in an insecticidal bioassay using larvae of the tobacco hornworm.

Given the persistence and the possible presence of exudates, there is potential for prolonged exposure of the microbial and invertebrate community to such toxins, and therefore studies should evaluate the effects of transgenic plants on both microbial and invertebrate communities and the ecological processes they mediate.

If transgenic crops substantially alter soil biota and affect processes such as soil organic matter decomposition and mineralization, this would be of serious concern to organic farmers and most poor farmers in the developing world.

These farmers cannot purchase or don't want to use expensive chemical fertilizers. They rely instead on local residues, organic matter and especially soil organisms for soil fertility (e.g., key invertebrate, fungal or bacterial species) which can be affected by the soil bound toxin.

Soil fertility could be dramatically reduced if crop leachates inhibit the activity of the soil biota and slow down natural rates of decomposition and nutrient release.

The long-term impacts of GMCs are not yet known.

GENERAL CONCLUSIONS AND RECOMMENDATIONS

The available independently generated scientific information suggests that

- the massive use of transgenic crops poses substantial potential risks from an ecological point of view
- the ecological effects are not limited to pest resistance and creation of new weeds or virus strains
- transgenic crops can produce environmental toxins that move through the food chain and also may end up in the soil and water affecting invertebrates and probably ecological processes such as nutrient cycling
- no one can really predict the long-term impacts that will result from such massive deployment of such crops.

Not enough research has been done to evaluate the environmental and health risks of transgenic crops, an unfortunate trend. Most scientists feel that such knowledge is crucial to have before biotechnological innovations are implemented. There is a clear need to further assess the severity, magnitude and scope of risks associated with the massive field deployment of transgenic crops. Much of the evaluation of risks must move beyond comparing GMC fields and conventionally managed systems to include alternative cropping systems featuring crop diversity and low-external input approaches. This will allow real risk/benefit analysis of transgenic crops in relation to known and effective alternatives.

The loss of agricultural diversity may lead to disaster in developing countries.

Moreover, the large-scale landscape homogenization with transgenic crops will exacerbate the ecological problems already associated with monoculture agriculture. Unquestioned expansion of this technology into developing countries may not be wise or desirable. There is strength in the agricultural diversity of many of these countries, and it should not be inhibited or reduced by extensive monoculture, especially when consequences of doing so results in serious social and environmental problems.

The repeated use of transgenic crops in an area may result in cumulative effects such as those resulting from the buildup of toxins in soils. For this reason, risk assessment studies not only have to be of an ecological nature in order to capture effects on ecosystem processes, but also of sufficient duration so that probable accumulative effects can be detected. The application of multiple diagnostic methods will provide the most sensitive and comprehensive assessment of the potential ecological impact of transgenic crops.

Conclusion: agricultural biotechnology is driven by profit, not by scientific research.

Although biotechnology is an important tool, at this point alternative solutions exist to address the problems that current GMCs, developed mostly by profit motives, are designed to solve. The dramatic positive effects of rotations, multiple cropping, and biological control on crop health, environmental quality and agricultural productivity have been confirmed repeatedly by scientific research. Biotechnology should be considered as one more tool that can be used, provided the ecological risks are investigated and deemed acceptable, in conjunction with a host of other approaches to move agriculture towards sustainability.

5

Agriculture Biotechnology: The Answer to Food Shortages in Northeast India

INTRODUCTION

The world economy has seen significant growth in the last decade through increases in productivity, product quality, and export base diversification. These advances were mainly driven by the growth of traditional agricultural and industrial sectors. The application of agriculture biotechnology offers the opportunity to alleviate marginal and subsistence farmers to get out of their dependency on government assistance and to increase production.

The introduction of new crop varieties with insect and herbicide resistant genes is quickly changing the landscape of agriculture in the US and many countries. Major pests can be eliminated with little or no spray of pesticides, as well as nasty weeds killed with the spray of herbicide in genetically engineered crop varieties. Genetically engineered crop varieties are environmental friendly with reduced chemical use as an added benefit.

Given the need to feed more people on the same land area while using less water and nutrients, the application of biotechnology to improve crop plants is fast becoming the only viable option for the teeming millions. Let me give a direct quote from Dr. Norman Borlaug, Noble Peace laureate, ?the world need to double food production by 2050 if hunger were to be banished from the world and the ongoing 'gene revolution' can definitely play a part in this. You cannot build peace on empty stomachs. Only 8 per cent of countries with lower levels of hunger are mired in conflict". I might add, insurgency, the dominant menace of the region is eating up the resources and turning many vibrant and intelligent youths into dangerous thugs because there is not enough opportunities to go around.

Many insurgency groups have sprang up as a result of hunger and a lack of opportunity rather than a genuine desire to leave the society in better shape than the one they currently inherit.

WHAT IS GMO?

The acronym GMO stands for ?genetically modified organism,? and refers to any plant or animal species that has had a gene or genes from a different species transferred into its genetic material via accepted genetic engineering techniques. The process of introducing genes into a new species and getting them to function is known as transformation. Genetically modified organisms have a foreign gene inserted into them that creates one or more new traits for that organism.

The genetic engineering employs an array of methodology such as gene discovery, transformation, molecular characterization of the insert, and gene expression study. In a real sense, ?genetically modified? is not a new phenomenon, but has been practiced by plant breeders for years through breeding and selection with the objective of producing more productive, more resistant offspring, or in producing better or different quality of product than the existing variety.

The process to produce new varieties following conventional breeding methodology takes approximately 6-7 years compared to 2-4 years via genetic engineering. A single gene with a known function can be moved from one organism to another without transferring additional unrelated genes, as happens in the process in conventional breeding method.

Above all, what marks GMO out from the products of conventional breeding was that the new techniques help us to cross taxonomic lines (across unrelated species), which means I can put a B.t. gene from a soil bacteria over to cotton.

HOW SAFE ARE GMO CROPS?

The majority of scientists have consensus that there is nothing inherently risky about genetically engineered crops. Splicing genes from one organism into another has been used for years. One example is the production of an enzyme required in processing cheese. No one calls that dangerous. Recently introduced genetically engineered rice (golden rice) that could potentially eliminate vitamin A deficiencies has been met with widespread acceptance.

Before a variety with a new trait is approved for cultivation, it has to go through rigorous government agency approval process. For example, in the US, three agencies: the Food and Drug Administration (FDA), United States Department of Agriculture (USDA), and Environmental Protection Agency (EPA) must approve the new gene(s) before the traits are released. The process is very involved and takes up to 2 years for final approval. The agencies ask the following questions: is it safe to consume (FDA); is it safe to grow (USDA); and is it safe for the environment (EPA).

THE BENEFITS OF GMO

The best way to demonstrate the benefits of GMO is through examples. 1) The virus that causes severe ring spot in papaya. The disease reduces papaya production and kills the trees in Asia, in parts of Latin America, and in Africa. 2) Leaf curls disease on white potatoes. 3) The leaf yellowing in sweet potatoes throughout east and central Africa. 4) The stunting and yellowing in rice throughout central Asia. Each of these examples is caused by different strains of virus. Each of these important diseases can be controlled through biotechnologies that increase the resistance of the plants to the viruses. The production of cotton in India, Pakistan, Egypt and other countries where the boll worm, boll weevil and other insect pests have in the past reduced yields, have with the application of biotechnology turned around and increased.

The common insect resistant trait gene used in the market place today is called B.t. gene. When smallholder farmers in China and South Africa grew native cotton varieties that contained the B.t. gene for insect resistance as a result of biotechnology, the farmers realized between $150 and $200 per hectare in increased profits. It is estimated that more than a million farmers (combined) in these two countries have benefited from insect resistant varieties of cotton. The increased profit came because of the reduced application of insecticides to control the pests and increased lint production.

Perhaps the most striking examples of how biotechnology can improve human nutrition are found in the many varieties of rice and canola that have been improved by biotechnology. Biotechnology has been used in rice and canola to increase the amounts of beta-carotene, the precursor of Vitamin A, which is in short supply in diets in many parts of the world. The hope is that consumption of foods rich in beta-carotene will alleviate the chronic Vitamin A deficiencies in the diets of many of the poor in Asia and Africa. Other research is underway to increase the levels of other vitamins, to improve the amount of proteins in crops like potatoes and cassava. Researchers are also developing foods that can deliver certain types of therapeutic substances, such as vaccines, which stimulate the body's immune system against certain endemic diseases.

BIOTECHNOLOGY CENTER IN NE INDIA

The NE region is rich in diversity-both culturally and agriculturally. With agriculture being the predominant industry of the region, any positive changes focused on here is welcome news for 70% of the population. Since land holding is limited, any increase in agriculture product must come from either changing the cultural practices or integration of agricultural biotechnology. The later is attractive and promising for the future since there is a limit to what cultural practices will add to the overall production. For example, the Green Revolution in the early 1960s with the introduction of dwarf varieties of wheat and rice depended on cultural practices to make the changes, but production leveled off after agricultural inputs where enacted.

Establishing a biotechnology center is imperative if the region is to improve the agriculture to meet the future food demands of the region. Crop improvement through genetic engineering must be a priority if the region wants to stay competitive and to best utilize natural resources. A number of states have already established their own biotechnology center as is in the state of

Andhra Pradesh. The region already has a number of universities that can provide the technical know-how and skills needed in the hundreds of competent faculty and scientists they employ. Additionally, there are several Indian Council of Agricultural Research centers in the region, which may have already thought through the process of establishing a biotechnology center.

For a biotechnology center to stay competitive, the center must reflect the following criteria:

(1) as a government (central and regional) sponsor with a non-political entity,
(2) governed by a board of directors (scientist should make up a majority of the board),
(3) registered as a non-profit organization,
(4) development process driven by product delivery rather than consumption driven,
(5) seek its own funding through government and foundation, both national and international,
(6) competitive funding scheme should ensure the integration of biotechnology with classical breeding and other associated technical disciplines,
(7) located close in proximity to a major airport,
(8) close to a major university or college,
(9) have the goal of becoming a self-supporting entity,
(10) have the objective of addressing regional issues,
(11) employee selection criteria strictly merit, and
(12) the seven sister states each contribute and play a role.

WHAT ARE THE ECONOMIC IMPACTS AND THE MARKET OUTLOOK FOR THE REGION WITH GMO?

The economic impact on the region is enormous-from food import dependency to a player in India and the world food market. The total area under GM crops globally has risen by 20 per cent in the last year alone. In India, where GM use was allowed only three years ago, about 1.34 million hectares have come under B.t. cotton. A cotton farmer in India can increase his bottom line even though he pays higher seed premium in order to grow genetically engineered cotton. The market outlook for GM products in India is excellent since food shortage and poverty are still alive in India.

In conclusion, I believe biotechnology can bring unprecedented economic impact in the region, which in turn will change the attitude of the people. I believe if a product and market driven economy takes a hold in the region, the young people will trade their guns for the newfound economic power they so desperately seek to achieve. Rather than stealing at gunpoint, they will work hard exploiting the new economic reality within their reach.

One final note, but economic freedom is only one of three freedoms I mention in an earlier article, ?True Freedom.? One cannot be truly free if he is not free spiritually. Spiritual freedom includes freedom from corruption, murder, envy, and all kinds of evil. Poverty and hunger in India can be eliminated, but the greater challenge is to eradicate the spiritual poverty that is beneath the surface of each person?s apparent happiness.

6

Advanced Agricultural Biotechnology

Every year, millions of people worldwide die from starvation and nutritional deficiencies. Although these challenges remain significant, we are making some progress. Agricultural technologies have increased caloric intake on a per-capita basis in nearly every nation of the world. technological development — particularly in the area of agricultural biotechnology — is crucial if we want to further reduce malnutrition and starvation and meet the needs of growing populations. Unfortunately, misperceptions about biotechnology have led to activist movements to slow down, if not eliminate, this critical technology via government regulation. In the United States, the advancement of biotech crops has slowed down because of attempts to regulate them differently than conventionally grown foods. In Europe, antibiotech movements are much stronger, and international bureaucrats are pushing various international agreements and trade restrictions. While Western nations may be able to absorb the costs of such policies, people in the developing world will pay a higher price as many continue to suffer from starvation.

MEANING

For centuries breeders have improved plants through selective breeding and crossbreeding different types of plants. Agricultural biotechnology — also referred to as "genetic engineering" or "genetic modification" (GM) — is a more sophisticated and precise means of modifying plant genes. Instead of transferring thousands of genes as traditional crossbreeding does, biotechnology enables the transfer of only a few select plant genes. Breeders essentially cut the desirable traits out of one plant's DNA and paste them into another — a process called "gene splicing." In addition, undesirable traits can be targeted and removed. Using biotechnology, plant breeders can produce better crops.

Biotechnology's Critical Benefits

More affordable, higher quality food surely benefits people in developed nations, but those benefits pale in comparison to what biotech promises for the developing world. Biotechnology promises to reduce world hunger and disease by improving local productivity by adapting crops to local climates and soils; increasing yield by making plants stronger and more pest-resistant; making plants more nutritious by creating plants with higher vitamin and protein content; and making produce more affordable on the world market.

OVERVIEW

For example, Cornell University researchers are developing a rice variety that yields 20 percent to 40 percent more food per acre and a tomato that produces 48 percent higher yields per acre. Biotechnology promises to play a critical role in reducing blindness caused by vitamin A deficiencies. Every year in Southeast Asia, one quarter of a million children go blind because of a lack of vitamin A in their diets. Worldwide, 124 million children suffer from vitamin A deficiency. Increasing the intake of vitamin A could prevent one to two million deaths per year and eliminate numerous health problems.

Poverty and lack of infrastructure make the traditional methods of vitamin consumption (a varied diet or supplements) impossible. Agricultural biotechnology, however, can help to address the problem of childhood blindness and death by ensuring that people get this nutrient in the key staple of their diet, rice. Researchers have spliced into the genome of rice, the beta-carotene gene from a daffodil to create "golden rice" — rice high in beta-carotene, which the human body converts into vitamin A. This is among the most well-known examples of biotechnology's benefits. Countless other genetically engineered crop varieties are being developed to address similar problems of poor nutrition or productivity.

How Agricultural Biotechnology Improves Plant Safety

Many biotechnology skeptics contend that such engineering is unsafe because it is "unnatural." Ironically, biotechnology is actually an improved method for plant breeding. Most of what we eat today has already undergone gene manipulation under the traditional breeding methods. The traditional methods and biotechnology differ in that traditional methods are cruder because they do not give researchers control over all the genes they crossbreed. Some desirable characteristics might result with such methods, but undesirable ones might as well. It could take a long time of continual breeding to eliminate the most problematic traits and focus on the best ones.

Compared to these more "traditional" methods of hybridization, genetic engineering is a wonder of precision and accuracy. Genetic engineering enables scientists to pinpoint the genes that express a desirable characteristic in one plant and to insert those genes into the genome of another plant. Scientists mark the insertion point so that they know where these novel genes are located in the genome of the new plant. While there are several different methods of genetically engineering plants,

AGRICULTURAL BIOTECHNOLOGY REGULATION

For traditionally bred plants, regulators rely on plant breeders to conduct appropriate safety testing and to be the first line of defense against genetic alterations that might prove dangerous. The same should be true for agricultural products developed using biotechnology. There is no reason for regulators to treat biotech plants any differently than traditionally bred plants, particularly given the fact that biotechnology provides greater control over gene manipulation. Despite fears about gene manipulation, traditional crossbreeding has altered plant genes and improved the human diet for the past 30 years. In fact, people worldwide safely consume these plants every day.

Examples include wheat, corn, potatoes, tomatoes, and countless other staples of the American diet. In 1984, the biotechnology industry began experimenting with "genespliced plants" — the more advanced approach to gene manipulation that we now call "biotechnology."

Regulatory Scheme

At the time, the White House Office of Science and Technology Policy began crafting a frame-work wherein existing federal agencies would regulate genetically engineered organisms on the basis of their characteristics, not the method of production — thus wisely deferring to the scientific community's judgment that regulation ought to address the products, not the process of biotechnology.

Engineered organisms would not require extra scrutiny simply because genetic engineering produced them (the process). Instead, they would be subject to heightened scrutiny only if the individual organisms expressed characteristics that posed some conceptually heightened risk (the product).

The federal government divided regulatory jurisdiction among agencies already involved in agricultural, food, and environmental regulation. These include the United States Department of Agriculture (USDA), the Environmental Protection Agency (EPA), and the Food and Drug Administration (FDA). While each of these agencies considers the characteristics of individual products in their regulation, only FDA followed the general scientific thinking that genetically engineered and non-genetically engineered products should be regulated similarly. Both USDA and EPA automatically subject all genetically engineered plants as a class to premarket approval requirements not ordinarily applied to conventionally bred plants.

Department of Agriculture

USDA regulates the release of all genetically engineered agricultural plants under statutes giving USDA's Animal and Plant Health Inspection Service (APHIS) the authority to regulate plants that may be or may become weeds or other nuisances — what the statutes call "plant pests." Although the rules apply in a general sense to novel or exotic varieties of both genespliced and conventional plants, APHIS typically only requires field testing of conventional plants that are new to a particular U.S. ecosystem (transplanted from another continent, for example).

However, all genetically engineered agricultural plants face a higher regulatory bar — making it more difficult to expand the use of biotechnology. Genetically engineered crops must be field tested under APHIS regulations prior to commercialization. For example, a new variety of corn produced with conventional hybridization requires no government-mandated field testing, but all new varieties of genetically engineered corn do, even though there is no logical reason for the regulatory disparity.

For most genetically engineered plants, APHIS requires the company producing the plants to submit notice detailing the gene or genes that have been inserted, where the plant will be tested, and other relevant characteristics of the plant prior to receiving permission to conduct the field trials. Once the company completes field testing, APHIS reviews the results and makes a determination on whether or not the product should be "deregulated" and can be released into the market.

Environmental Protection Agency

EPA regulates plants that are modified to produce substances that act like pesticides: that is, substances used by a plant to protect itself from pests, such as insects, viruses, and fungi. EPA's

proposed rule for these "plant-incorporated protectants" is not yet finalized. In the interim, however, EPA has regulated such plants by applying its proposed guidelines, which are functionally similar to rules for the registration of synthetic chemical pesticides. Again, biotech products face higher regulatory hurdles, even though plant-incorporated protectants developed through conventional breeding are exempt from these requirements.

Health and Human Services: Food and Drug Administration

FDA is responsible for ensuring that food items, including foods derived from genetically modified plants, are safe to eat. Following the general regulatory framework that emphasizes product regulation rather than process regulation, FDA rightly does not treat foods derived from genetically engineered plants to be inherently unsafe.

Food producers are not required to seek premarket approval from FDA unless there is a substantive reason to believe that the novel trait (or traits) in the food pose a safety question.

The initial determination of safety is left to the producer, but FDA has encouraged producers to consult with agency scientists prior to marketing a food produced with biotechnology to ensure that the appropriate determination is made. Recently, FDA published a proposed rule that would require producers to notify the agency at least 120 days before marketing, which may be a signal that FDA could abandon its more reasonable approach in the future.

In addition, FDA does not require labeling of foods derived from biotechnology unless the genetic insertions so alter the food that the common name no longer describes it adequately. Examples of this phenomenon would include such alterations of the food product that would raise a safety issue or change the product's nutritional composition or its storage or preparation characteristics.

Labeling Issues

Some activists, however, argue that the government should mandate the labeling of all genetically engineered foods. They assert that consumers have a "right to know" how their foods have been altered, and that a mandatory label would best allow consumers to choose between genetically engineered and conventional foods.

Biotechnology advocates have argued against mandatory labeling because such requirements raise food costs — something that mostly harms lower-income Americans and people on fixed budgets. Perhaps more important, while biotech products are not substantially different from other products, special labels would likely make consumers think these products were more dangerous.

Hence, rather than serving educational or "right to know" purposes, such labels promise to simply confuse consumers.

A government-mandated label on all genetically engineered foods also would raise important First Amendment free speech issues. In 1996, the U.S. Court of Appeals, in *International Dairy Foods Association et al. v. Amestoy*, ruled unconstitutional a Vermont statute requiring the labeling of dairy products derived from cows treated with a genetically engineered growth hormone, noting that food labeling cannot be mandated simply because some people would like to have the information. "Absent ... some indication that this information bears on a reasonable concern for human health or safety or some other sufficiently substantial governmental concern, the manufacturers cannot be compelled to disclose it."

In other words, to be constitutional, labeling mandates must be based in science and confined to requiring disclosure of information that is relevant to health or nutrition. Furthermore, consumers need not rely on mandatory labeling of biotech foods to truly have a choice. Realworld examples show that market forces are fully capable of supplying information about process attributes (including kosher and organic production standards) that consumers truly demand. The same can be said about nonbiotech foods, and the FDA recently published proposed guidelines to assist producers in voluntarily labeling both genetically engineered and nongenetically engineered foods. Additionally, the USDA's newly published rule for organic certification necessarily excludes biotech products from organic food production.

Consequently, consumers wishing to purchase nonbiotech foods need only look for certified organic products.

International Trade

While U.S. consumers do not appear to be strongly opposed to biotech foods (they, in fact, seem rather indifferent), a strong anti-biotechnology movement has arisen in several European countries over the last few years. The European Union (EU) has established strong restrictions on the commercial planting of genetically engineered plants, and European food processors and retailers are reluctant to import harvested agricultural products derived from biotechnology — thus jeopardizing the marketability of U.S. commodity grain exports. And the EU is now negotiating for strong restrictions on agricultural biotech products in important international agreements governing food safety and environmental protection. Very strong restrictions were included in the Biosafety Protocol, finalized in January 2000.

Perhaps, more important, though, are the Codex Alimentarius Commission standards for food safety. Without convincing scientific evidence that genetically engineered crop plants pose a heightened environmental or human health risk, restrictions on agricultural biotech imports could be challenged under the General Agreement on Tariffs and Trade and the World Trade Organization. Thus the EU has resorted to justifying its decisions on the basis of a risk-management philosophy known as the "precautionary principle," and the EU advocates inclusion of the precautionary principle in international environmental and food safety agreements (such as the Codex Alimentarius), as well as within GATT itself. Its inclusion would give nations greater latitude in restricting imports of U.S. agricultural products.

No single definition exists for the precautionary principle, but its general meaning is that when an activity raises threats of harm to human health or the environment, regulatory measures should be taken to prevent or restrict the activity even if the risks have not been demonstrated scientifically. Although the EU asserts that the precautionary principle is an unbiased risk-management philosophy, critics have noted that its lack of definition and evidentiary standards makes it all too easy to abuse for the purpose of masking trade protectionism, and that its very approach to risk management is inherently flawed and may, in fact, increase net risk.. Nevertheless, the precautionary principle has already been included in several international environmental treaties, including the Convention on Biological Diversity and the Biosafety Protocol.

Even as underdeveloped nations clamor for biotechnology applications, and as countries like China continue to experiment with and use agricultural biotechnology, opponents of agricultural biotechnology in the West, particularly Europe, attack it as an unnatural process that will destroy the world, not better it. They argue that biotechnology should be heavily regulated, if not banned. Already, however, genetically engineered plants are subject to strict regulatory oversight that is equal to or greater than that advocated by the vast majority of scientific specialists. Additional regulation will slow down research and development of genetically engineered crops, keep beneficial products off the market, and raise the cost of products that do make it to consumers. Furthermore, the inclusion of similar restrictions — or inclusion of the precautionary principle — in international agreements will greatly impact the international trade of agricultural goods and delay their introduction into the marketplace. Each of these problems could prevent this technology's benefits from being introduced to industrialized nations and, more important, the developing world.

TWENTY YEARS OF MODERN AGRICULTURAL BIOTECHNOLOGY

In 1973, Cohen and Boyer transferred a gene from one organism into another. In 1982, the first biotech plant, an antibiotic resistant tobacco, was developed. In January 1983, at a meeting of genetic researchers in Miami, three different teams reported success in using Agrobacterium tumefaciens, a bacterium, to carry new genes into plant cells, heralding the dawn of modern agricultural biotechnology.

Agrobacterium tumefaciens, described as a "natural genetic engineer", splices its own genes into host plant cells. This pathogenic bacterium was now converted into a pack mule, to carry new, foreign genes into plant cells, minus the disease and this became the most common means of producing Genetically Engineered Organisms (GEOs).

Field tests for GE crops resistant to pests and pathogens were first conducted in the US, in 1985. Accoordinated framework for the regulation of GEOs was established and the first GE tobacco was released, in 1986. The US Department of Agriculture published guidelines for field trials of GE crops in 1991.

On approval from the Food and Drug Administration, Flavor Saver, the first GE tomato, with a longer shelf life, was on the US markets in 1994. During 1995-96, GE soybean, corn and cotton were approved for commercialization, in the US.

A number of GE crops, developed for pest, pathogen and herbicide resistance, are now commercially cultivated in several countries. Rice with pro-vitamin A, higher iron content, or human milk proteins and potatoes with high protein content are in various stages of development. Tobacco plants producing functional human haemoglobin and bacteria that produce human insulin have been developed, as well as plants with vaccines against rabies and other viral diseases. Food grain crops that withstand drought and salinity are high priority research, and so are those for high yield. A GE tobacco plant detoxifies soils contaminated by explosive residues, providing solution to a frustrating environmental problem in countries ravaged by armed conflict.

Now more than 70 biotech agricultural crops that have been approved for use in North America, including varieties of soybeans, cotton, canola, corn, potatoes, squash, tomatoes and papaya. About six million farmers in some 17 countries now cultivate GE crop on about 125 million acres, a 30 fold increase over 1996. By end of the year 2002, six GE crops planted in the US (soybeans, corn, cotton, papaya, squash and canola) produced an additional four billion pounds of food and fibre on the same acreage, improved farm income by US $ 1.5 billion and reduced pesticide use by 46 million pounds. In 2003 in the US, 80 per cent of soybean acres will be planted with biotech varieties.

A number of activist groups vehemently attack GE technology and its products on grounds of safety to humans and the environment, and costs of technology transfer and its reach to the needy. Products of agricultural biotechnology bear the brunt of this ill-informed, unscientific and prejudiced onslaught much more than GE products related to health or industry.

In 2001, the European Community released results of a 15-year study, costing US $ 64 million, and involving more than 400 research teams and 81 projects. This report concluded that GE products pose no more risk to human health or the environment than conventional crops. So far, extensive and intensive research on the probable risks of GE technology has not brought out any adverse effects and none of the fears expressed by anti-tech activists were proved even marginally. Nevertheless, caution and examining issues case by case, is the watch word of technologists, who are aware of their responsibilities.

For a number of products, technology transfer is free of costs for developing countries, as for example Golden Rice, the rice with pro-vitamin A. It is the responsibility of the Governments of the respective countries to bear the costs of developing local varieties and to reach the products to the needy at an affordable cost.

In India, three GE varieties of pest resistant cotton were approved for commercialization, a year ago. Approval for other GE varieties of cotton and GE mustard was deferred twice by the

Genetic Engineering Approval Committee (GEAC), the highest authority on the issue.

In India, the level of public awareness of the realizable benefits and probable risks of GE products is abysmally low. The functioning of the GEAC leaves much to be desired. Taking advantage of this hazy situation, mixing up economic, social and political issues with science, and even using such vagaries of nature as the severe drought, several groups of vested interest have created mistrust, confusion and scare. Trashing a very promising technology this way does not augur well for the future of Indian agriculture. This results only in denying the benefits of technology to the farmers and consumers.

The next 20 years of plant biotechnology is expected to improve quality of life, through improved foods, pharmaceuticals, new industrial materials and a better environment.

These benefits should reach the people of the developing world, who need them most. For India to benefit from these developments, it is urgent and essential that the public is educated on the realizable benefits and probable risks related to GE products. We also need to reorganize the mechanism of regulating GE products by providing for an expedient, transparent and responsible authority. The media have a very important role to play in this regard. Only an informed and reassured public can make viable choices. Every effort should be made to provide such an opportunity to the consumer.

MANAGING AGRICULTURAL BIOTECHNOLOGY IN COLOMBIA

The international scenario for biotechnology shows a rapid tendency at industrialized countries in the increase of publications, patents, enterprises and novel solutions for the industry, the environment, health and agriculture. Nevertheless, Colombia has an important delay in relation to the international scientific development and the capacity to generate wealth and services for its productive systems.

This delay has been one of the concerns of the country's policy during the last years, and more precisely since 2002, when for the first time biotechnology was included in a National Development

Plan as one of the mechanisms for competitiveness and the use of biodiversity and genetic resources.

This paper is the result of a survey conducted in 2005 aimed to provide an overview of agrobiotechnology in Colombia to be included in the compendium of case studies organized by the FAO's Regional Office for Latin America and the Caribbean (LAC) and the Network for Technical Cooperation in Agricultural Biotechnology in Latin America and the Caribbean (REDBIO/ FAO).

The command of skills on classic biotechnologies in Colombia has allowed the strengthening of capacities, especially with micropropagation and cell and tissue culture. Nevertheless, the transition towards modern biotechnologies has been difficult, especially due to the scarcity of financial resources and the access to new technologies and specialized knowledge. Neither has the country a critical mass of economists, managers and lawyers to integrally support the processes of technological supervision, competitive intelligence, market access, national regulations and international treaties on intellectual property rights and biosafety.

The attention that necessities of small producers deserve, together with the improvement of the intellectual property rights, public perception and biosafety framework on genetically modified organisms (GMOs) for human and animal consumption, are some of the fields on which we should be working in the country through international networks, with an aim to reach the benefits that use of modern biotechnology brings along, as some researchers in this field have highlighted (Watal, 2000; Spielman et al. 2006). In this way, the current technological gap existing between countries of the South and the North could be bridged (Huete-Peìrez et al. 2001; Byerlee and Fischer, 2002; Tollens et al. 2004).

GENERAL BIOTECHNOLOGY INDICATORS

Since 1999, organizations such as the Organization for Economic Co-operation and Development (OECD) and the Organization of American States (OAS) began to compile statistics to describe the situation of the countries that are developing biotechnologies. A general view shows that countries in Latin America and the Caribbean (LAC) have significant gaps with countries of the North, especially in patterns of patenting and licensing technologies, and that the transferal of biotechnologies is of vital importance, especially those biotechnologies developed for the agricultural sector (Testa, 2002; Heisey et al. 2005; Thumm, 2005; Chan, 2006).

In LAC countries, measurements mainly include the analysis of the capacity of groups and research centers, their working fields, the number of enterprises and the market perspectives, whereas the statistics of OECD measure mainly the private investment, the generation of investor's capital in terms of new companies, money and employment, and the obtainment of patents among other indicators, which are not so representative for us due to the differences in economies and development systems.

An indicator to analyze the state of biotechnology in the countries is a function of the investment made for its developments and products. The United States of America has a much higher investment tendency in relation to other countries. In 1997, the industry invested around US$ 7.300 millions, being 3.7% of the total investment for research and development (NSF, 2004), and as for 2001 it reported that investment of industry in research and development of biotechnology reached US$ 16.400 millions, representing 10% of the total investments in science and technology (U.S. Department of Commerce, 2003). European countries such as Germany, with more than US$ 1.000 millions (6.7% of total), United Kingdom with US$ 705 millions (7.8% of total) and France with US$ 560 millions having a 4.4% of the total of governmental investment in R&D, follow the list with the largest assigned budget for biotechnology according to statistics of OECD (Van Beuzekom, 2001).

The situation in countries members of the OECD reveals a striking difference between USA and EU. According to data provided in 2001 by the consultant company Ernst and Young, USA investment in R&D was € 11.400 million (2.7% of DGP) whilst EU investment was € 4.977 million (1,8% of DGP). There are 1,262 companies in USA totalling 162,000 employees and 6.7% of them are involved in research, in contrast to 1,570 companies in EU with 61,000 employees, 2.5% of them conducting research (Ernst and Young, 2001).

In the compendium of biotechnology statistics of OECD, differences can also be observed. The first one is the amount of investment in risk capitals in biotechnology, which goes up to US$ 1.182 millions in the USA, and US$ 687 millions for the EU during 1999.

The bioengineering products reached that year up to 1% of the total exports of USA with a total amount of US$ 1.340 millions. Its import of products from OECD countries, especially from Belgium, France, Switzerland and The Netherlands were that same year of US$ 970 millions. Its most important commercial partners where most of the exports are sent to include: Belgium, Japan, Canada and Germany (Van Beuzekom, 2001).

This indicator allows analyzing the degree of technological delay resulting from the efforts of countries such as Colombia, where it represents approximately US$ 1.8 millions (constant Colombian pesos as in 2003.

Exchange rate 1 US$ = $ 2.200 Col) and 3.7% of the total governmental investment of the National Biotechnology Program (NBP) made through Colciencias during 1997, and during the following years the tendency remained negative, reaching in 1999 almost US$ 780.000 and 4.8% (OCyT, 2004). This situation somehow explains why Colombia reports 8 patents granted to research groups associated with the NBP (OCyT, 2004), whereas the USA already had more than 3.600 patents and Canada more than 500 during the year 2000 (Van Beuzekom, 2001). It could be said that the relation between the investment in biotechnology and the number of patents is more than US$ 2 million per patent in the USA, US$ 640.000 in Canada and US$ 22.000 in Colombia.

On the other hand, the development of biotechnology in LAC has been characterized by being a repetitive model where the research projects do not correspond to the real needs of production and food security, and has a low level of innovation and scientific creativity. Only 2% out of 30 scientific articles of high impact on biotechnology came from laboratories based within the region (Izquierdo and De La Riva, 2000), despite LAC possessing important biodiversity resources to develop innovative products (Roca et al. 2004).

THE STATE OF BIOTECHNOLOGY IN LATIN AMERICA AND THE CARIBBEAN

Bourinbaiar (2006) recently reviewed developments achieved outside EU that were presented at the BIO 2005 meeting and highlighted progresses done in Argentina and Chile among LAC countries. A comprehensive review was undertaken based on five important books published between 2003 and 2005 by multilateral organizations, showing comparisons and experiences among several LAC countries in terms of scientific and technological capacity, political situation, biosafety, commerce, biodiversity and international markets.

The first study, made by CamBioTec in 14 Latin American countries, Argentina, Bolivia, Brazil, Chile, Colombia, Costa Rica, Cuba, Ecuador, Guatemala, Mexico, Peru, Paraguay, Uruguay and Venezuela, clearly shows that for 2002 the biotechnological industry was represented by 432 companies (including less than 20 companies in the agricultural sector in countries of the Andean Community of Nations (CAN)), where the most outstanding are the BioMinas cluster in Brazil, the agro biotechnological pole of the National Institute of Agricultural Technology (INTA) from Argentina and the Western Biotechnological Pole (Polo Biotecnológico del Oeste) in Cuba. The companies mostly involved with the agricultural sector are in the fields of cell biology and microbiology, with emphasis on the propagation of plant materials to lower cost and time production. Regarding national regulations for the protection of intellectual property rights, microorganisms can be patented in Brazil and Mexico, while in Chile it is possible to patent biotechnological processes (CamBioTec, 2003).

From the second study, financed by Iberoamerican Program for the Development of Science and Technology (CYTED) during 2003, it can be concluded that while the majority of researchers work in the agricultural sector, innovations are not so representative when compared to the pharmaceutical sector. It also shows that one of the topics where the policies of the countries (with the exception of Brazil) have had less result is in the creation of new enterprises with a biotechnological base. This is due to the shortage of monetary resources and the lack of continuity in the national programs and plans (REVYDET, 2003).

The third one, published in 2003 by OAS, CamBioTec, the USA Embassy in Chile and the Chilean Government, shows the results on different topics related to the application of the Cartagena Protocol and foods derived from genetically modified plants. The study shows experiences that allow us to have an overview of the situation, the perspectives and challenges that modern biotechnology brings to agriculture and it pretends to facilitate the discussion on the risks and benefits of biotechnology for the region (OAS et al. 2003).

The fourth study was financed by OAS and published by the Colombian Observatory of Science and Technology (OCyT), presents statistics from Colombia, Costa Rica, Mexico and Venezuela together with the comments from different experts on the construction of indicators in biotechnology. In the different comparisons made with individuals dedicated to R&D, research facilities, training programs, projects and products, there is an evident lack of univocal usage of the notion of biotechnology, while in some countries activities of second generation are accounted, and in other countries only the projects involving the use molecular biology and genetic engineering are accounted (OAS and OCyT, 2004).

Finally, during 2005 the Andean Development Corporation (CAF) produced an analysis of the markets and capacities of the five countries of the CAN, to make use of biodiversity. From the CAF's study it can be highlighted that from the 567 research groups from the five participant countries, half of them work for the agricultural sector. According to the indicators developed by Dr. William Roca for the CAF, the Andean region has reached a 52% of development of the scientific and technological capacity considered optimal to make use of biotechnology and biodiversity. Regarding the financial tendencies, the study reports that from the 22 groups selected from the five countries that get yearly funds for research, they get on average US$ 97.500, within a range from US$ 30.000 to US$ 1.5 millions (CAF, 2005)

In the Network for Technical Cooperation in Agricultural Biotechnology in Latin America and the Caribbean (REDBIO/FAO) there are four case studies at the national level for Argentina, Bolivia, Ecuador and Peru available at www.redbio.org/

estud_casos.htm. According to these studies, we can see in the case of Argentina that modern biotechnology has had an enormous influence from multinational companies controlling the market of transgenic seeds, and that there has been important progress on surface area cultivated with GMOs (Diamante and Izquierdo, 2004). Bolivia shows very little progress in biotechnology, especially on laboratories involved in research, enterprises involved in agricultural produce marketing and as licenses for access to genetic resources (Avila and Izquierdo, 2006). Ecuador, on the other side, has had some progress in the research on Andean roots and tubers and on tree tomato but nevertheless, its progress have been very limited, especially with the introduction of GMOs and the use of biodiversity and genetic resources (Wendt and Izquierdo, 2003). Finally Peru has only one request for introducing a GMO from the International Potato Center (CIP), and its progress at the institutional and political levels has not allowed yet the strengthening of the capacities needed for the effective use of the potential represented in biotechnology (Pastor, 2004).

About the current capacity of the LAC, the data bank CATBIO from the Network for Technical Cooperation in Agricultural Biotechnology in Latin America and the Caribbean (REDBIO/FAO), founded in 1991, provides information on 738 laboratories located in governmental, private and academic institutions from 32 countries in the region, and from 4300 professionals working in areas related to research, regulation, extension, education and diffusion of biotechnology, all of them integrated in the module *profREDBIO*.

BIOSAFETY: PROGRESS IN LEGISLATION

Social, economic and ecological implications when adopting biotechnologies and regulatory frameworks on biosafety have been subject to analysis especially when dealing with agricultural applications of LAC countries (Ramírez, 2003). However, the weak inherent regulatory policies have promoted more scientific progress rather than industrial applications in the LAC region (Solleiro and Gálvez, 2002)

Assessments of potential risks that GMOs could generate for human health, food and the environment are undertaken as part of biosafety studies. The risk assessment should be made on the properties of the product, independently of the techniques used for its development. The risks should be evaluated in relation to the characteristics of the transferred gene and the properties of the plant or receptive organism, and the specific ecosystem where the transgenic will be introduced. Another accepted principle, especially in European countries, is the so called Cautionary Principle, on which the inexistence of evidence on the potential harms is not a reason not to make the standards considered necessary to prevent their occurrence (Torres, 1999). Studies have been carried recently on pollen flow from Bt maize (Peterson et al. 2006) and flow of genes conferring resistance to insects, diseases and herbicides mainly with potato, tomato, eggplant, cabbage, pepper and cucumber have been subject to risk assessment (Ram and Dasgupta, 2004) especially in Europe, where concern on GMOs is evident (Van Den Eede et al. 2004).

The Cartagena Protocol on Biosafety is the regulatory framework subscribed by more than 100 countries in the world and states the agreements for implementing common procedures on biosafety as well as the rules and regulations for safe trans-boundary movement of GMOs discussed mostly in the Article 18, Paragraph 2, with the aim of conserving biodiversity and environment (Pythoud, 2004; Verastegui et al. 2004).

In addition, the activities related to the development and implementation of the national standards, in the frame of the Global Environmental Facility and World Bank (GEF-WB) Project: "Capacity development for implementing the Cartagena Protocol on Biosafety –Convention on Biological Diversity in Colombia" aimed to the inter institutional agreement and articulation of the normative and adaptation of each of the points considered in the Cartagena Protocol. The GEF-WB project has been a scenario of interaction among the participant institutions in search of the articulation and implementation of the national normative under effect, related with the implementation of the Cartagena Protocol that has impacted decisively on the articulation of institutional capacities in terms of normative frameworks that allow a better functioning of the dynamics of innovation and new markets (Verastegui et al. 2004)

During 2000, Colombia signed the Cartagena Protocol on Biosafety to the Convention on Biological Diversity, a legally binding instrument that enters into force in our country by means of the Act 740 of 2002, implying the creation of a biosafety system for the country (OAS and OcyT, 2004). Within the context of the Cartagena Protocol, specifically in the items related to the implementation of rules and regulations on biosafety in Colombia, the country already has an acting legislation in the items of uses and applications of GMOs and its derivatives for agricultural, animal and human consumption. Also are defined the competent authorities, the Ministry for Agriculture and Rural Development (MADR), Ministry of Environment, Housing and Territorial Development (MAVDT), the Ministry for Social Protection (MPS), and indirectly the Ministry of Commerce, Industry and Tourism (MCIT) and the Ministry of Foreign Affairs (National Chancellery) as the national focal point for the Cartagena Protocol.

In the Ordeal 4525 of 2005, three National Technical Committees on Biosafety were created as organizations with the power to examine and evaluate the submitted applications, advising on the issuing of administrative acts that authorize the development of activities with GMOs. This ordeal applies to the transboundary movement, transit, manipulation and use of transgenic organisms, and basically divides the competency of national authorities, leaving the responsibility to Colciencias, in accordance with the use of GMOs, which accompanies the MADR and the Colombian Institute for Agriculture (ICA) in aspects related to agriculture, livestock, fisheries, forestry and agro industrial processes, the MAVDT in relation to the environment; and the MPS and National Institute for the Surveillance of Drugs and Food (INVIMA) in relation to human health and feeds.

For food evaluation, the Specialized Court for Food of the INVIMA was created, entity affiliated to the MPS, which by the Ordeal 936 of may 1996 which makes the evaluations that are scientifically pertinent for the approval process for the use in the country of additives, of new products developed by biotechnology and the other products of the competence of the court. Additionally, the ordeal 3075 of 1997 in its article 54 establishes that will be special procedures for food obtained through third generation biotechnologies and/or processes of genetic engineering, with the purpose of awarding the sanitary registry, previous study and favorable concept from the Revising Commission-Specialized Court for Food, according to the Ordeal 936 of 1996.

The MPS by Resolution 00485 of March 2005 establishes the technical regulations on the requirements for labeling or tagging that must be accomplished by canned food and raw materials of food for human consumption. This resolution is currently under discussion and will enter in force by the end of 2005. Article 5 of the resolution specifies the information that food labels and tags must bear. Numeral 5.2.2 states that it must be declared on "any food or food ingredient obtained by means of biotechnology, the presence of any allergenic transferred from any of the products listed in the literal paragraph of the present article. When it is not possible to provide adequate information on the presence of any allergenic through the label, the food that contains the allergenic could not be commercialized". According to article 10, "irradiated food or food placed under ionizing radiation and the ones obtained through certain techniques of genetic modification or genetic engineering, must accomplish with the specific regulations for labeling that for its effect the National Government issues". Currently, the government is developing this disposition for the labeling of GMOs for human consumption.

The competent authorities designated have studied the applications submitted for approving the use and commercialization of GMOs and their products, which involve studies on risk assessment, management and monitoring. Since 1998 the CTN of Agricultural Biosafety ICA has received and processed 12 applications, out of which 9 have been approved, each one with its specifications as a commercial crop, field trials under contention and confined research. Regarding the studies on risk assessment and the introduction into Colombia of GMOs, between 2000 and 2001 the CTN started the evaluation of the Bollgard® technology for the case of cotton owned by Monsanto; it is the first case in the country. In this evaluation two studies were made. From the study of pollen flow it was concluded that gene transfer by means of wind dissemination is statistically null (ICA, 2002). A second study concluded that the Bollgard®

technology had very significant effects on the pest targets, without affecting the beneficial insects such as useful arachnids needed in the pollination of the cotton flower (ICA, 2003).

PUBLIC PERCEPTION

As new types of transgenic organisms, vaccines and pharmaceutical products are developed and continually introduced into markets, biosafety becomes the main focus of public concern (Wang, 2006). There have been controversies on GMOs release and consumption where mass media have played the major role. Consumer acceptance of GMOs is complex and diverse across cultures (Blaine et al. 2002; McInerney et al. 2004; Bauer, 2005). Countries such as USA, with an export tradition of soybean, canola and maize, developed GM crops to enhance their competitiveness. Regulation issued by the US Food and Drug Administration (FDA) related to production and labelling is objective and expeditious, while the EU has created a complex regulation for GMOs (Vidal, 2003; Fernández-Díez and Corripio Gil-Delgado, 2004; Herrick, 2005).

Between 1998 and 1999, EU invested an annual average near to US$ 100 million on agrobiotechnologies (Kalaitzandonakes, 2000), at the same time as the application of rules and regulations on labelling for GMOs trade was initiated, which was rejected by large food trading companies as Carrefour and Marks & Spencer and European citizens in overall (Vidal, 2003; Durant and Legge, 2005). This situation resulted in a significant reduction on soybean and maize imports from USA and in a divided public opinion in Europe, rendering GMOs object of a controversy among consumers.

McCluskey and Swinnen (2004) state that 90% of consumers receive information on food and biotechnology via press and television, but there is not any comprehensive information body to ensure the creation of a risk public perception. For instance, 45% in Canada (Einsiedel, 2000) and 30.1% in the USA (Hornig, 2000) think that biotechnology is risky. In the EU 63% of the citizens think that decisions must be based on moral and ethical considerations. On the other side, those who believe that we have the right to exploit nature for human wellbeing (70%), are prone to think that biotechnology and genetic engineering could have a positive effect on life quality for the next twenty years (Durant and Legge, 2005; Eurobarometer, 2005).

Colombia is aware of the debate between the movements that are trying to return to the ancestral agriculture by promoting through green labels and certificates the identification of organic products, and the movements that promote the consumption of transgenic food that provide benefits for the producers, consumers and the environment.

The public perception on biotechnology was analyzed in Colombia through two different sources of information. The first one, understood as the formal speech, transmitted through the media and documents released by the governmental institutions, which make opinion, inform, take decisions and rule the activities that make use of biotechnology. The second one, based on oral information, includes the research groups, education institutions, national and multinational companies, consumer associations and environmental movements, that use speeches that fundament the realities in the public sphere. These types of studies have also been conducted in Europe by means of surveys in order to open the public debate on GMOs in the way as Pidgeon et al. (2005) did in Great Britain.

During 2004 Agro-Bio gathered press information to study the situation on the positions towards the GMOs. Of the 47 opinion articles analyzed, 54% reveal a positive position, while the remaining 46% is divided in equal parts between negative and neutral. From the most representative titles, we highlighted in the newspaper El Colombiano: "Cotton: giant recovering", (Feb. 23), "Genetically modified organisms-GMOs", (April 26); in the newspaper: Portafolio: "Biotechnology and development", (Feb. 25) and "The debate on transgenics", (Oct. 17); from El Universal we highlight: "The century of biotechnology", (Jun. 8); in La República, "Insects and Bt" (Sep.6); and finally from the newspaper El Tiempo we highlight: "Biodiversity is not patentable" (Jun. 22) and "What's on with biotechnology in Colombia?, by that time head of the National Program on Biotechnology of Colciencias, and an article written by Alfonso López Michelsen, former President of Colombia and a key person in the political sphere at the national level, and published in the opinion column from the November

21, 2004 edition, entitled "Positive results in Colombia: Agrobiotechnology in 2004", where he gives his point of view on the benefits that the country has received thanks to policies promoting the adoption of GM crops, especially the progress in cotton.

Regarding the second source of information, one of the cases is the result of a research done by the Universidad del Rosario that allows to conclude that the overall knowledge of the citizens is little, and that from the semantic point of view, interviewees associate biotechnology with food that has been altered, manipulated and transformed, and that have a connection to laboratories working on genetics, where abnormal, artificial, good, improved, natural and preserved produce are obtained (Perales, 2001). Another study conducted by ICA during 2002 for the city of Bogotá, shows that the attitude towards the consumption of GM food is 42% positive and 12% indifferent.

The perception on the "chemicals" and the "transgenic crops" is currently generalized in terms of the alteration of the natural composition of food, and the associated impression of risks to health that are brought along their consumption (Perales, 2001). The election of a certain organic produce, a transgenic produce, or one produced by conventional methods is tightly related to different social groups. While in the upper strata groups the decision making is based on a wide spectrum of criteria, in the lower strata the decisions are related to their limited economical capacity, where the price is the main criterion. The former could get access to organic produce, which have in the country a restricted offer due to only 1% of farmed lands with them and only 10% of their production destined to the national markets, while the latter must get the more accessible produce according to their income.

Globalization, the debate on food security and potential risks of biotechnology, as well as strong pressures from interest groups such as multinational companies and environmentalists, all demand starting education programmes similar to the United Nations University Biotechnology Programme for LAC, to promote an open dialogue among scientists, opinion leaders and mass media, that would act as the basis to support the technological shift during economic development (Braun, 2002).

In order to generate an objective public perception on transgenics, it is necessary to have suitable information sources showing scientific evidence. For this, there have been activities such as workshops, meetings, seminars and congresses at the national and international level, aimed to different sectors of the society. Some of the main groups that these activities are aimed to are government officers, policy makers, media, professional associations, scholars and the productive sector. During these meetings, several topics such as the use and development of biotechnologies, biosafety, risk management and assessment of GMOs, and regulatory frameworks are discussed (Juanillo, 2001; Verastegui et al. 2004).

INTELLECTUAL PROPERTY RIGHTS

Intellectual property is a legal system that provides exclusivity rights to persons and companies for the commercial exploitation of their creations and inventions. The granting of Intellectual Property Rights (IPRs) has been justified to assure the return of the inversion that is made to develop an invention. This is especially critical when dealing with new technologies, where biotechnology has a protagonist role.

Brazil and other LAC countries have an immature national innovation system where scholar and scientific progress receive more support than industrial developments do, particularly due to the construction process of a normative framework on intellectual property (Coutinho et al. 2003). The protection of intellectual property rights in agricultural and pharmacy biotechnology is also the cornerstone of the dispute about bio-piracy. The international protection of intellectual property has been a contentious issue between developed and developing countries (Kerr et al. 1999; Forero-Pineda, 2006).

The first patents in modern biotechnologies appeared during the 80's, when the patent to recombinant DNA technology was granted in the USA to Stanley Cohen and Herbert Boyer, which has given more than 200 licenses to pharmaceutical and biotechnological companies and received more than US$ 100 million in royalties. Pray and Naseem (2005) show how patents on genetic transformation techniques stimulate private investment and reveal how benefits derived from patenting exceed the costs. However, there have been conflicts regarding the protection of intellectual property rights on genetic transformation technologies, especially in the case of viral vectors, *Agrobacterium*, and other bacteria, among governmental institutions, universities and the private sector (Chung et al. 2006).

It is important to mention that at the World Intellectual Property Organization (WIPO), a conflict arose in 2000 between the objectives of the Convention on Biological Diversity (CBD) and the Legal Aspects of Intellectual Property Relating to Trade (ADPIC), WTO-(Aspects des droits de propriété intellectuelle qui touchent au commerce) around the discussion on the Patent Law Treaty (PLT), that led to the creation of an Intergovernmental Committee on Genetic Resources, Traditional Knowledge and Intellectual Property.

Also in the Committee on Trade and Environment (CTE) and the ADPIC Council of the WTO there have been discussions, especially referring to the article 27.3 b) that discusses two aspects of patentability. The first one excludes patents on plants and animals, including those genetically modified, as well as animal breeds and plant varieties. The second one includes the obligations of the countries to guarantee the protection of microorganisms, non-biological processes, microbiological processes, and plant varieties through either patents or an effective *sui generis* system, or a combination of both (Vivas, 2001).

The majority of innovations on plant biotechnology are subject to be patented, even that the coverage range varies considerably according to the country and the technology. The patents are used to protect biotechnological tools, reagents, genetic sequences, and transformation, regeneration and diagnosis processes. The obtainment of a GMO includes the insertion of a construct containing a gene of interest, a selection marker gene, a promoter, and other sequences that could be patented.

The insertion can be achieved by means of a transformation method that can be patented if it has novelties when compared to the original method, whether using equipment or any other material for which an exception can be done with a research activity. Obtaining a patent on a new gene does not guarantee the freedom to operate when other technologies or processes could be used for the gene of interest to work in the plant.

The regime of patents affects biotechnology from its use in plant improvement by means of hybridization to the development of transgenic plants. The technologies for the diagnosis and the equipment and procedures used to study the genomes are susceptible of patenting. Nevertheless, when a new variety could not be a novelty or protected genetic material, if this development is related with the use of techniques or processes, it could be subject to licensing or restriction for plant breeder's rights.

In the same way, and according to what is stated in article 20 of the Decision 486 of the CAN, it is considered that inventions whose commercial exploitation on the territory of the respective member country makes an attempt against public order, moral, health or life of people or animals, or the preservation of plants or the environment, are not patentable. Finally, are not subject to patenting, plants or animals and essentially biological procedures for the production of plants and animals that are not non-biological nor microbiological procedures, neither the therapeutic or surgical processes for the treatment of humans or animals, neither the methods of diagnosis applied to human beings or animals.

In Colombia the CAN disposition is ruled by the Decree 2591 from December 13, 2000 and the regulating resolution 210 from January 15, 2001. The IPRs are managed by the Surveillance of Industry and Commerce (SIC), adhered to the MCIT, which in turn is in charge of giving the patents and registration of marks. By means of Law 243 of 1995 plant breeder's rights were established and in 1996 the country adhered to the act of International Union for the Protection of New Varieties of Plants (UPOV) of 1978. The legal framework on IPRs is basically given by the Andean decisions 345 for the UPOV and 486 for industrial property. Within the analysis it should be taken in account what was agreed in the ADPIC agreement and the world context. It should be especially considered the aspects related with genetic material, derived products and the non tangible component associated with them.

Plant breeder's rights have a more limited coverage than patents because they only protect the variety, but not the obtainment procedures nor its uses. For a variety to be subject to protection it must have the following characteristics: novelty, which means it has never been commercialized; distinctiveness, which can be differentiated clearly from other known varieties, uniformity and stability. The time coverage of the protection is 15 years for short-cycle crops and 20 years for permanent crops.

According to SIC, plants and genes are not susceptible of patenting in Colombia, but non-plant GMOs and specific processes made by microorganisms can both be patented. That means the biotechnological processes that imply the use of biological material for the production or transformation of products as is the case of fermentation, are patentable. Equally are patentable the procedures to obtain transgenic plants, a plant cell that has been transformed, nevertheless, the complete transgenic plant is not, neither any of its reproductive structures. Modified genes, recombinant DNA, cDNA, vectors that hold genes and do not exist as such in nature, and cells with foreign genes are all patentable in Colombia.

Finally, the challenges on intellectual property derive from how to harmonize the purposes of science, industry and society: a scientific development that generates basic knowledge, with an application beneficial for improving life quality for humankind; an industry that makes innovations and has a compensation system for its investment; and a vigilant society receptive to innovations that are at its reach and constitute a source of social and economical well-being.

ACCESS TO GENETIC RESOURCES

The United Nations Environment Programme (UNEP) promoted the creation of the CBD, signed by 150 government leaders in 1992. "The Convention recognizes that biological diversity is about more than plants, animals and micro organisms and their ecosystems-it is about people and our need for food security, medicines, fresh air and water, shelter, and a clean and healthy environment in which to live" (CBD, 1992). One of the main commitments contained in the CBD, which was ratified by Colombia by means of the Act 165 of 1994, is the one referred to the creation by each country of conditions enabling access and conservation of genetic resources. The CBD states in its objectives that an adequate access to genetic resources should be allowed and appropriate transference of technology required for investigating these resources should be facilitated.

- Colombia recognized the Andean Decision 391 of 1996 to establish a common regime of access to genetic resources, understanding for genetic resources "any material of biological nature containing genetic information of actual or potential utility or value" and for access "obtainment and utilization of genetic resources conserved *in situ* or *ex situ*, their derived products, or-whether the case-their intangible components, with the purpose of research, biological prospecting, conservation, industrial application or commercial use."

Colombian flora is legally protected by means of the Act 299 of 1996 and botanic gardens are regulated as scientifically organized collections of living plants. Botanic gardens are enabled to manage herbaria and plant germplasm in the form of genebanks (or seedbanks), shall conduct permanent programmes on basic research, education, and *ex situ* and *in situ* conservation. One of their major goals is genetic diversity conservation. In addition, Article 16 defines the activities to be carried out by the Colombian National Herbarium-Natural History Museum of the Instituto de Ciencias Naturales, Universidad Nacional de Colombia, the herbarium of the Instituto de Investigación de Recursos Biológicos Alexander von Humboldt along with all the other official herbaria, as well as those affiliated to the Colombian Association of Herbaria.

Besides, Decree 1603 of 1994 orders the Instituto de Investigación de Recursos Biológicos Alexander von Humboldt to keep gathered the biological collections resulting from scientific plant collecting permits and those obtained by regional environmental agencies and other research institutes ascribed to the MAVDT. The Instituto de Investigación de Recursos Biológicos Alexander von Humboldt within a work leaded by Ricardo Torres in 2004, identified 25 germplasm banks, 13 botanic gardens and 36 herbaria, totalling 22,750 germplasm accessions, 14,490 accessions container in botanic gardens, and 1,344,000 exsiccates.

Worthy to note is the fact that the country exhibits the highest biodiversity index (0.935) when compared to Mexico (0.928) and Brazil (0.877), rendering Colombia a megadiverse country *par excellence* (Roca et al. 2004). Despite, the country had some difficulties to make good use of this advantage. Calle (1996) described some of the technical, legal and sociocultural difficulties that Colombia have to face on the access to genetic resqurces, and Zerda-Sarmiento and Forero-Pineda (2002) described some aspects in the relationship between large firms and traditional knowledge in ethnic groups in some LAC countries that will be taken into account.

CONCLUDING REMARKS

It is well known that one of the potentials that Colombia has in the agricultural ambit is on its large biodiversity and variety of climatic levels and altitudinal ranges, but its sustainable use will occur only after the potential species and its market niches have been positively identified, to permit the country to be competitive at the national and international markets, solving problems especially for food and pharmacopoeia. The production and marketing of agricultural products and natural resources demands today, more than ever, certifications, guarantees of plant health, resistance to biotic and abiotic factors, innocuousness and highly productive performance, and also the development of productive chains allowing to hold demands in time. This is how the exposed cases in Colombia allow seeing the way on how networks were created and consolidated among heterogeneous actors, where science and technology achieved their contribution for the inclusion of small producers in the dynamics of global competition with the use of biotechnology (Orozco and Chavarro, 2006). Nevertheless, there is still much to be done, especially in the sustainable articulation of productive chains and the creation of competitive sustainable advantages making use of biotechnology as a mechanism for innovation.

The present study identifies three key aspects for agricultural biotechnology development in Colombia: (1) a substantial investment increase, which should be focused on the obtainment of patentable industrial applications utilizing biodiversity and genetic resources; (2) an effort increase of governmental and non governmental institutions in implementing and applying regulations on biosafety, intellectual property rights and access to genetic resources, by stimulating investment and innovation; and

(3) providing opportunities for open discussion and wide dissemination of information that will favour building an objective public perception and thus progressing towards a decision-making capacity of consumers and the general public.

It is also clear that the only way to overcome many of the limiting factors that have been faced by the potential use and development offered by biotechnology in a country like Colombia, is through a state policy that overcomes the changes generated in each government, and this is why the policy proposal Visión Colombia II Centenario 2019, which contains a chapter entitled "Making use of biotechnology, biodiversity and genetic resources" is the opportunity to achieve biotechnology to be projected in the long term and be inserted into the productive systems and environmental sustainability, generating impacts especially for small producers, understood as the increment of their possibilities to get access to a better life quality, the enjoyment of wellbeing and to expand their possibilities of freedom in the process of human development. The key point of this policy is to generate added value to biodiversity and promote the industrial development of products, emphasizing on patent obtainment by research groups and companies.

Colombia requires a multisectorial institution a) facilitating the incorporation of benefits derived from biotechnology in productive systems by enabling biotechnology transference and linkage, b) conducting a permanent monitoring of activities, their impact and inclusion into various sectors of application, and c) providing advisory on intellectual property, technological transference, biosafety, public perception, access to genetic resources and markets. It is likewise required from such an institution the coordination of policy design and implementation based upon prospective surveys, exercises on technological surveillance, market surveys and sociocultural analyses.

The information compiled through interviews, information systems and bibliography sources allowed us to make an evaluation of the generation, application, implementation and use of agricultural biotechnology by different social actors that through policies, programs, cooperation networks and transference activities, have made possible to incorporate biotechnology into some agricultural productive systems in the country (Schuler and Orozco, 2005).

The study cases developed by many LAC countries, as an initiative of REDBIO/FAO, will allow to know the enormous potential that these countries count with, to bring together the generating actors of science and technology, so that through an associated work, supervised by technological assistance, LAC may represent an integrated model of technological development, not only for agricultural biotechnology, but also for the other areas into which there is applicability such as health, environment and industry. This study is inscribed into the general frame of contributions that have been produced with the aim to analyze the strategy that integrates science, technology and society, as the engine for the development of nations.

7

Agricultural Biotechnology Sector in India: Issues Impacting Innovations

INTRODUCTION

Agricultural biotechnology sector in India, particularly its plant biotechnology segment, is uniquely poised for a major process of transformation. As the country's agricultural biotechnology sector made strides in the 1990s in r-DNA, transgenics and molecular marker assisted plant breeding process, the Government of India responded with a matching policy support and regulatory framework that was designed to render the path of progress in R&D, sustainable and bio-safe. To a large extent, developments in the policy front have been induced by a vibrant non-governmental sector that intensely intervened on the sensitivities of modern biotechnology. This paper discusses analysis of the issues and proceeds to further consider the path of progress achieved by agricultural biotechnology companies in India in the area of commercialization of biotechnology products. The special focus of this paper is on plant biotechnology.

The biotechnology sector in India in 2002-03 is estimated to have attracted private investments to the tune of US$ 140 million. Infrastructure and R&D account for nearly 80 per cent of these investments.

During the period from 1986 to 2002 it is estimated that the Government of India invested an amount of US$ 275 million in the biotechnology sector. The Government of India has also extended venture capital support for biotechnology start-ups through the Technology Development Board and the New Millennium Indian Technology Industry Leadership Initiative (NMITLI). Meanwhile the States of Andhra Pradesh, Karnataka, Maharashtra, Gujarat, Kerala and Tamil Nadu have drawn up ambitious sub-national biotechnology policies, which promise fiscal and infrastructure support measures to prospective entrepreneurs. Some State Governments have launched their own venture funds for supporting novel start-up ventures.

To catalyse the process, the Government of India accorded priority to IPR protection. The Indian Patents Act of 1970 was amended to provide for product patents for agro-chemicals, drugs and pharmaceuticals and micro-organisms over a transitory period. A product patent regime would be in place in India by January 1, 2005. The new patent law extends protection term for inventions from 14 to 20 years. To provide for protection of IPRs in respect of new plant varieties, Parliament has enacted a *sui-generis* legislation in the shape of the Plant Varieties Protection and Farmers Rights (PVFR) Act in 2001. These developments have created favourable legal conditions for international partnerships in biotechnology R&D. By prioritizing FDI investments in biotechnology, the Government of India has sent appropriate signals to international investors. Further the fact that these changes have been in consonance with the WTO-TRIPs has lent greater creditability to these changes.

Meanwhile, the Government has also brought in regulations over agricultural biotechnology. Apart from the biosafety regulations that form part of the Environment (Protection) Act 1986, the Indian Parliament has legislated the National Biodiversity Act in the year 2002, to provide for regulatory controls over access to biological resources in India.

IPR REGIMES FOR PLANT BIOTECHNOLOGY

Article 27(3) of the TRIPs Agreement lays down that members shall provide for the protection of plant varieties either by patents or by an effective *sui-generis* system or by any combination thereof. Article 27(3) in particular is critical to India as this requires us to provide protection of plant varieties by conferring plant breeders rights either by way of patents or through *sui-generis* plant variety protection legislations. Countries with strong R&D base in plant genetic engineering such as USA have robust Plant Utility Patent Legislations.

India is certainly not inclined to adopt patent protection regimes for its plant varieties. India is rather inclined to adopt a *sui-generis* legislation which is non-'patent' based. There are many reasons for this proclivity. India is one of the ten mega-diversity countries of the world and a rich storehouse of landraces of principal agricultural crops. India has a strong R&D base in conventional plant breeding methods. Its strength in plant genetic engineering is impressive but not an overwhelming factor by comparison. The first two strengths explain India's disinclination towards plant patent regimes or towards a *sui-generis* legislation which is 'patent'-driven.

While Plant Utility Patents Act provide for broad patents over plant varieties, traits and genes and even the physical parts of the plants, plant breeders rights provide for IPR only over varieties. As is well known, since 1990s the UPOV (Union for the Protection of Varieties) has been largely viewed by developing countries as offering the best regime for positioning their national legislations.

The central feature of the UPOV is the protection it affords to plant breeders who produce plant varieties that fulfil the criteria of distinctiveness uniformity and stability (DUS). The current version of the UPOV, viz. UPOV 1991 had added additional criteria of 'new' to DUS thus rendering DUS as NDUS. Contrary to the popular notion, the NDUS criteria of UPOV 1991 is not substantively different from the principles of 'novelty', 'inventiveness' and 'industrial application' (NII) which applies for patents. The criteria of 'novelty' and 'inventiveness' in Patent Laws are covered by the criteria of 'new' and 'distinct' in UPOV 1991. Thus by distinctness, the UPOV means a variety of plant which is 'clearly distinguishable from other varieties whose existence is a matter of common knowledge'. It is apparent that this term captures the attributes of 'novelty' and 'inventiveness' implicit in Patent Laws. Even in respect of 'uniformity' and 'stability' criteria the UPOV does not offer different recipes. True, by 'stability', the UPOV conveys that 'relevant characteristics of protected plant variety remain unchanged either for a specified period or after repeated propagations or cycles of propagations'. It is also true that 'stability' is a difficult criteria for a plant breeder to fulfil. Attainment of 'stability' criterion is problematic for cross-pollinated plants and non-single homozygous lines of autogamous plant varieties. This, in turn, reduces the commercialisation potential of the plant variety since inconsistency of genetic quality jeopardises commercial application of the plant variety. The same holds true of the criterion of "uniformity". Therefore, the NDUS criteria of UPOV 1991 are homologous to the NII criteria implicit in Plant Utility Patent Laws.

The UPOV has undergone two major amendments since its inception in 1961, i.e. in 1978 and 1991. The main concern with regard to currently operational UPOV 1991 has been its stringent provisions, which include the additional attribute of 'new' to the DUS criteria. The scope of UPOV 1991 in terms of coverage is larger. While plant varieties of all taxa are covered by UPOV 1991 only plant varieties of nationally designated species were covered by UPOV 1961 and 1978. Further by defining clearly that 'variety' represents a 'group of plants', or 'single' or 'several plants' or 'one' or 'several parts' of a plant(s) the UPOV 1991 has considerably enlarged the scope of its coverage. Going by this definition even sexual and vegetatively propagated varieties are covered by UPOV 1991.

The US PVPA legislation has been quick to adopt the UPOV 1991 formula. The amendments of 1994 to the US, PVPA broadens the term 'plant varieties' to include all materials harvested from the protected varieties. The other drastic feature of UPOV 1991 has been abridgement of the 'plant breeders exemptions' and 'farmers privileges' conferred by the earlier versions. Earlier versions of UPOV allowed farmers to re-use seeds generated from their farms for self-use, and allowed breeders to freely use protected varieties for further improvements. These concessions gave UPOV a 'less stringent' face in comparison to 'patent' laws. UPOV 1991 gives an option to National Governments to disallow the farmers privileges of retaining or re-using seeds for self cultivation. UPOV 1991 also restricts the breeders rights by enlarging the right of the breeder to close variations of his cultivar or 'essentially derived varieties'. In other words, the breeder of a cosmetically bred variety would have to buy genetic dependency rights from the derived variety prior to commercialising the derived version. Further the duration of protection have also been extended under UPOV 1991 to 20 years for crops and to 25 years for trees and vines. Thus in terms of these rigorous features, UPOV 1991 eliminates the arbitrage between plant varieties protection and patent regimes.

IPRS AND INVESTMENTS IN PLANT BIOTECHNOLOGY

The implicit rationale for IPR protection is that they promote investments in plant breeding and bio-engineering. This hypothesis is contested by section of analysts. Nevertheless it is a fact that the structure of IPRs do influence investments. IPRs induce their own pattern of innovations. The time taken for R&D to fructify as innovations is a crucial aspect guiding investments.

Often IPR regimes play a key role in influencing the time pattern of innovations. It is a well-known fact that conventional plant breeding methods are dilatory and time consuming. It requires 7 to 8 generations of repeated breeding to reduce heterozygosity of new genotypes. As Vossen (1985) observes Arabica Coffee subject to three to four cycles of breeding still display heterozygosity and it is not until 20 to 25 years (spanning 4 to 5 generations) that one arrives at true breeding varieties. This time lag is even more pronounced for the Arabusta coffee variety (inter-specific hybrid of Arabica and Robusta Coffee) where segregation of unfavourable characteristics could still occur even after several generations of backcrossing and selfing.

Another interesting example is the one cited by Evenson (1991) about the long gestation period in regard to rice varieties in India. According to Evenson, India released 306 rice varieties for planting during the period 1965-1986 after performing 20,000 crosses over a period of over 15 years since 1950.

It must be understood that this long process of backcrossing was preceded by considerable time spent on location and identification of individuals plants with requisite traits. The speed or pace at which plant varieties qualify for being released for planting have economic implications on investment behaviour in seed/plant breeding industry. Asexual modes of plant breeding which include plant genetic 'engineering' and transgenic techniques could obviate the long gestation period associated with pre-release preparations encountered conventional breeding.

It is a different matter than biosafety concerns may delay releases of these varieties. The time lag in release of plant varieties fulfilling the criteria of novelty, distinctness, uniformity and stability (or alternatively the requirements of novelty, inventiveness and commercial application) raise fundamental issues from the view-point of capital investments. The time lag in varietal release is potentially least in the case of transgenic and non-sexually produced varieties while for traditional or classical methods the time lag can be greater.

The UPOV 1991 based plant IPR regimes could induce longer time lags as compared to the situation when an IPR regime is based on the 1978 version of the UPOV. This is due to the fact that the minimum genetic distance concept implicit in UPOV 1991 precludes 'close cousin' varieties from acquiring protection rights, unless the techniques of genetic distance determination by themselves are inadequate.

UPOV 1978 would facilitate capital investments to some extent as the release of plant varieties is periodically more frequent. Even under the plant patent regimes, investments could bring about short-term returns, though the volume of investments will have to be larger on account of capital intensity of advanced plant genetic engineering technologies.

By comparison, UPOV 1991 regime is least conducive to capital investments as plant variety releases can be painfully slow under this regime. Therefore sustenance of conventional plant breeding under UPOV 1991 regimes, can only be ensured by infusion of low discount capital.

Paradoxically it is the reduced availability of low discount public funds for plant breeding that has induced countries like India to go in for Plant Varieties Protection Legislations modelled on UPOV lines. The implicit assumption in undertaking this step is that private funds for R&D in plant breeding will 'crowd in', if protection plant varieties rights are accorded. But the reality is that in liberalised deregulated developing economies, private investments have 'crowded in', in short return promising avenues.

In other words, private investment funds for R&D are not likely to be channellised into medium to long return yielding areas.

Let ra, rb & rc be the discount rates relevant for regimes A, B & C in Figure I. Then the relative rates of discount will be r1 > rb < rc and relative present worth will be pa < pb > pc.

Consequently classical plant breeding activities will suffer for want of funds in the UPOV 1991. In other words, UPOV 1991 modelled national legislations will not necessarily hold attraction for private sector investors in R&D projects involving classical plant breeding as compared to say, UPOV 1978 or the Plant Patent regimes. Rather private investments would be catalysed better by *sui-generis* plant protection legislations, which partake of plant patent elements and focus in the main on hybrids and genetically engineered varieties.

Given the linkages between plant breeding and the seed industry these perverse investment behaviour trends can produce major economic consequences. The Seed Laws in various countries, both advanced and developing, are getting to be progressively oriented towards restricting market circulation to varieties that are conferred with plant breeder or patent rights. The UPOV guidelines on DUS are used world wide not only as the basis for establishing varietal distinctiveness and descriptions but also for seed certification purposes. The DUS guidelines are employed for recognising and registering not only the 'basic' seeds used for multiplication of 'pedigree' seeds but also for certification of the pedigree seeds themselves.

Evenson (1991) explains how absence of IPR rights will reduce Marginal Variable Costs (MVC) of seed multiplication. Tougher controls by certification of seeds could ensure that seed supply is kept in a 'constricted mode' and MVC of seed multiplication are raised. This in turn, will produce its own supply-demand dynamics as *Figure I* depicts. Figure I brings out the implications of IPR regimes in terms of the supply curves for seeds. As Figure I shows that in a situation of 'liberal' plant varieties protection laws (which do not provide for genetic distance) and equally liberal 'seed laws' which permit non-registered varieties (including landraces) to be sold in open markets, the macro-supply curve of seeds will be elastic (S1). When these regimes shift towards 'genetic dependency' PBR regimes and tighter seed laws that prohibit marketing of non-registered varieties, the supply curve will tend to be relatively inelastic (S2).

Analogously, demand curves could shift from D1 (wherein demand for protected varieties will be elastic on account of availability of landraces and non-protected varieties in the markets) to D2 (where the demand for the protected variety will be inelastic on account of ban on sale of landraces and other non-protected varieties).

In the D1-S1 situation, supply constraints and demand flexibility could induce low price regimes while in D2-S2 the reverse situation would occur. D2-S2 situation is not desirable for countries like India for other reasons as well.

The experience of transitional economies of East Europe, which had re-modelled their Seed Laws in conformity with European Community Seed Laws, has been sad. Farmers of the East European bloc cultivating and conserving landraces have been adversely affected by new seed laws, which by prohibiting trade in non-registered varieties not fulfilling the DUS criteria have contributed to the erosion of landraces in these countries. It is thus clear that agro-biodiversity can be adversely affected by the uniformity criteria of DUS, which brings about resultant changes in demand and supply position of seeds.

POST R&D REGULATIONS AND INVESTMENTS

Post-R&D regulations play a critical role in deciding on the pace of investments in the biotechnology sector. Even if a start-up enterprise enjoys a favourable IPR regime, that rewards its inventions, it is unlikely that it will invest in product development, in case post-R&D regulations are rigid and dilatory in nature. This then forms the backdrop for strategic alliances between start-up ventures and downstream industrial complexes. Multi-layered regulations in the post-R&D phase involve complexities which are expressed in the delays in commercial application /utilization of invented products or technologies.

In the case of third generation plant biotechnologies, such as transgenics, the problem is compounded by the fact that regulations by themselves are complex and less understood by the regulators themselves. A transgenic plant would have to undergo multi-stage regulatory checks and clearances before finding commercial application.

India is uniquely situated as far as biotechnology investments and infrastructure are concerned. In the area of plant biotechnology, advances have been made in transgenics of rice and wheat carrying stress tolerant genes such as CodA, COR47 and HVA1. The larger challenge is to take these products through the regulatory processes. More formidable is the issue of market development and appropriate pricing of transgenic seeds.

There are many imponderables here. Market shares for seeds in India have varied from crop to crop and from region to region. The 'demand' for transgenic plants or seeds may be even more uncertain, given the 'safety' dimensions and risk aversion propensity of Indian farmers.

In the dry agricultural pockets of Central and South India, farmers tend to be attached to traditional drought-hardy varieties of plants, which in turn contributes to their initial 'inertia' to shift to transgenics. Thus the task of marketing and pricing of novel biotechnology plants gets to be complex in such environment of 'demand incoherence'. A high margin pricing strategy for seeds may choke demand for transgenics.

Thus the degree of risks in commercializing transgenic plants could be of a high order. Given these realities, the idea of strategic alliances between start-ups and downstream entities looks to be realistic. The fact that premier R&D institutions such as the Indian Institute of Science, has set up its own start-up biotechnology venture, renders a strategy of tie-ups, a logical option. Much will depend upon the complexities inherent in product development and the willingness of competent downstream entities to signup agreements. To a large extent, this will depend upon enabling policies at the upper end of the value chain.

To sum up, India's agricultural biotechnology industry has the necessary policy and regulatory support to promote active R&D. However, in the critical area of product development and commercialization much more attention needs to be paid. The multi-tiered regulatory framework that exists could seriously affect the performance of start-up companies that would like to run through the entire life-cycle of biotechnology product development. Consequently, partnerships and strategic alliances should be encouraged between start-up ventures and established companies that have high product development capabilities. The experience of other countries in the Asia Pacific region could be a key input in framing development strategies for biotechnology products in India.

AGRICULTURAL BIOTECHNOLOGY

Since time began we humans have always relied on plants and animals to provide us with food, shelter, clothing and fuel, and for thousands of years farmers have strived to better these in order to continue providing us and to meet the needs of evolution. As the population grows and expands then so does the need for resources provided by plants and animals.

With the population expected to reach over 10 billion by the year 2030 it is estimated that the world food production will have to double on the farmland that exists today, if it is to compete with the anticipated growth in the population. The ever-increasing needs of yields of crops and the decreasing crop inputs such as water and fertiliser can all be helped by biotechnology as can advancements in providing pest control that is better for our environment.

CROP BIOTECHNOLOGY

Horticulturists and farmers have always relied for centuries on cross breeding and hybridizing plants and making genetic modifications to them in order to make improvements in the output and quality of food and fibre crops. They have made great advancements in building in protection against pests and diseases and protection from the harsher elements.

We learned from the Stone Age farmers to make a selection for next years crops from those plants that were strongest and healthiest and to save seed from these with which to replant. By selectively sowing seeds the earliest agriculturists performed modifications genetically to turn plants growing in the wild into domesticated crops, a long time before the science of genetics was really understood.

Over the years, knowledge and understanding of plant genetics improved and we gradually began to purposefully cross breed plants with the traits we wanted in order to produce plants which had the benefits of both parents, leaving undesirable traits behind. Virtually every plant that is grown for food today has come from cross breeding or hybridization or a combination of both.

Although we have come a long way in such a short time the process we were using to produce these crops was time consuming, costly and inefficient. Here is where the tools of biotechnology come into play, biotechnology allows us to take single genes from plants that give us the desired traits and move them freely from one plant to another.

This is a far more precise process than what we used to use and eliminates the thousands of genes of unknown function which was transferred along with the good traits into our crops. Biotechnology can also help us to remove the technical obstacles when moving genetic traits between plants and opens up a whole new world in benefiting food production.

HOW BIOTECHNOLOGY CAN IMPROVE CROP PRODUCTION

Agricultural scientists are incorporating biotechnology into crop production and protection, using the same traits that have been used throughout the decades and incorporate cross breeding and genetic modifications. Techniques are being used to increase the size of the yield of the crops, make crops resistant to disease caused by bacteria, fungi and viruses and give them the ability to better withstand harsher elements.

Biotechnology has also given us a better understanding of how we can make better use of the plants built in defence systems and it has opened up new avenues allowing us to work with nature by providing us with new bio pesticides.

These new pesticides allow us target crop pests but do not harm humans, animals, birds, fish and insects; they are also beneficial as they can control pests which have become resistant to the more conventional pesticides.

THE NEXT REVOLUTION IN AGRICULTURE

India is one of the largest agrarian economy in the world. However, the agriculture output has declined in last one decade because of various reasons like disease, pests and weeds, the subsistence nature of farming, unprecedented climatic conditions, limited water, poor land condition, drought, heat and saline soil conditions.

India has 25% of the world's cotton acres, an estimated 4 million cotton farmers, and some 60 million people depend on cotton and the textile industry to make a living. Textile industry accounts for India's single largest export sector. However the country produces only 12% of the world's cotton. Its average yields are among the lowest in the world.

There is thus a need to improve the techniques of producing cotton. Integration of biotechnology can accelerate the pace of agricultural improvement and is relevant to an agrarian country like India.

Crop biotechnology can provide the needed fuel and can be the engine of growth. It can help reduce poverty, promote rural development, strengthen trade and economy and agricultural sustainability, and also deliver direct benefits to farmers and consumers.

India has witnessed a tremendous rise in cotton production since the introduction of Bt cotton technology in 2002.

Monsanto all know what value Bt cotton is bringing to farmers in India, and the challenges faced by Bt cotton technology in the country. The year 2006 has been a year of successes and challenges for crop biotechnology in India.

The company prides itself to have been associated with Indian Agriculture for more than 55 years now. But it is also a fact that Monsanto and its Indian partners are facing one of the toughest challenges.

So much has been written about it in the media now that many of you know by now that tne company are sadly contesting an unprecedented interim order by the Monopolies and Restrictive Trade Practices (MRTP) Commission. While the MRTPC order has directed us to revise the fee of our technology in line with what we charge in China, the governments of Andhra Pradesh and several other states have sadly gone one step further in determining the maximum selling price of Bt. Seed, irrespective of the technology provider, to Rs 750. This is when the Indian farmers have hailed the commercial production of genetically modified cotton, welcoming it as a miracle solution for hard-hit cotton growers. Indian farmers are growing Bt cotton on an ever-increasing area because it delivers consistent benefits in terms of reduced pesticide use and increased income.

I would like to give some basic details about Bt cotton:

Bt cotton is grown by more farmers than any other crop biotechnology product in the world today. India is just one of the nine countries where Monsanto and its partners sell Bollgard cotton technology. Bollgard creates significant benefits for farmers in every one of these countries.

1) It replaces the costs of spraying pesticides
2) It increases farmers' yields.
3) It brings other benefits in terms of human and environmental safety, efficiency and peace of mind Bt cotton also a need to address the larger food and environmental issues and has reduced the use of pesticides and risks to the health of cotton farmers while improving both yields and farmers' incomes.

Therefore, if I were to put it in a nut shell, our basic philosophy to price our technology is to share most of the value that is generated with the farmer. But these benefits vary from country to country, because farming is different in each country. In some countries pesticides may cost more. In other countries, the yield benefit from Bollgard cotton may be more dramatic.

In spite of this variability, there is one common denominator farmers everywhere use to measure the economic value of Bollgard cotton-profit per acre. Because Monsanto's business is based on making farmers more profitable, the company price Bollgard technology based on the value the technology delivers in rupees per acre. When the company price Bollgard cotton, it consider all the variables in a given market. Once the company understand how much additional profit, yield, time and other benefits a farmer can expect to gain per acre from our technology, it set a price that ensures the farmer takes home a substantial part of Bollgard's value as his own profit per acre.

The company keep some of that value to share with our seed company partners, to cover our costs of doing business, and to reinvest in developing future products for farmers, including exciting new technologies such as drought tolerance. Many, who question the difference in prices of Bollgard technology between markets in different countries, have done so in the name of fairness to farmers. Ironically, some of these arguments about Bollgard price forget the way farmers make more money.

8

The Current and Future Outlook of Agricultural Biotechnology

INTRODUCTION

During the past few years, biotechnology has emerged as a powerful technology that has a big potential for a number of economic sectors, including agriculture. A number of basic human needs such as food, cloth, medicines and a cleaner environment have been fulfilled through biotechnological means. New developments in genomics (namely the human genome project and other genome projects in a number of organisms including important crops such as rice), bioinformatics and proteomics have opened new frontiers in biotechnology R&D. The convergence of these new technology platforms with the powerful information technology now provides a greater understanding of the biology of plants, animals, humans, and micro-organisms at the molecular level, enabling us to develop better strategies and approaches to deal with current and future challenges. The last millennium witnessed the shift of the Malaysian economy from its reliance on the raw and unprocessed products to manufactured goods. However, agriculture continues to play an important role in the overall economic growth of the country.

During that period, the Malaysian agricultural sector was left with two major challenges. The first major challenges is in the realm of addressing national food security, i.e. to produce sufficient amount of food to meet the national needs. The country is not producing sufficient amount of food for the populace. A drastic change in the exchange rate of the ringgit during the economic crisis of the late 1990s reinforced the above need as imported food items tend to be more costly. Of the total RM32 billion in the year 2000 budget, about RM10.52 billion was allocated for the imported food bill.

The second major challenge relates to creating wealth for the nation through production of value-added food and food products, which are more competitive in the open market, and to supporting the manufacturing sector through production of sufficient amount of raw material. The far-sighted vision on the role of the agricultural sector in supporting economic growth of the country requires crucial injection of new high technology to transform this sector to be more productive especially in meeting the two challenges stated above. The agricultural sector must prepare itself from now not only to produce sufficient amount of food, but it should also produce high quality products that can compete with the imported produce during the current era of globalization. The National Agricultural Policy outlined a transformation route for the agricultural sector to be more productive, efficient, competitive and more commercialized.

The last millennium also left us with the technology that will enable us to address the above challenges. The discovery of DNA and rapid development in the science and technology of molecular biology during the last millenium has ushered us into a new era – the biotechnology era. During that period, biotechnology has proved its ability to transform a sector – such as the agriculture sector – while generating new industries to support economic growth in the developed world. Genes, the basic biological unit of life, that programme a character of plants, animal and micro-organisms, are amenable to modification, manipulation and transfer across species to bring about desired outcomes. Biotechnology has been recognized as one of the new high technologies that will bring about desired changes in the agricultural sector.

INSTITUTIONAL DEVELOPMENT

Historically agricultural biotechnology development in Malaysia can be divided into three phases: the period of establishment (prior to 1995), development phase (1995-2000) and advancement phase (beyond 2001). Prior to 1995, initial establishment of agricultural biotechnology R&D took place in a number of research institutions such as MARDI, PORIM (now MPOB), and RRIM (now MRB) as well as at various universities such as UPM, UKM, UM and USM. The development at the institutions includes establishment of basic infrastructures such as molecular and cellular biology labs with necessary equipment and set-up, and establishment of basic expertise to undertake biotechnology R&D. The National Working Group on Biotechnology was set up under the Ministry of Science, Technology and Environment (MOSTE) to oversee and coordinate biotechnology activities in the country.

Subsequent to the successful establishment of basic infrastructures and technical competencies, many developments have taken place during the period of the 7 th Malaysia Plan (1996-2000). Implementation of the national initiative and agenda on biotechnology was further enhanced with the establishment of the National Biotechnology Directorate (BIOTEK) under MOSTE. The mission of BIOTEK is to spearhead the development of biotechnology in Malaysia through research and related activities directed at commercializing biotechnology, and to establish Malaysia as a leading centre for biotechnology industry. Under BIOTEK, a mechanism called Biotechnology Cooperative Centers (BCC) was established to assist the Directorate in coordinating the National Programme in Biotechnology; developing a network among participating institutions (universities, research institutions and industries); and pursuing leading edge research and accelerating diffusion of knowledge to the industry.

Seven BCCs have been established to cover primary biotechnology research areas, viz. plant-, animal-, food-, environmental/industrial, medical-biotechnology, molecular biology and bio-pharmacy. Besides the BCC system, another mechanism called the Contact Group Programme was also established by BIOTEK to facilitate direct communication and linkage with participating institutions of the public and the private sectors.

Recently, the Malaysian Government has endorsed a mega plan to develop biotechnology in the country under the BioValley Malaysia. The plan is based on the cluster concept where groups of specialized companies that support each other to create a "center of excellence". Companies within the cluster can take advantage of the presence of physical infrastructures, facilities, human resource, entrepreneurship and sharing of ideas to enable them to compete at a global level.

R&D

During the 7 th Malaysia Plan (1996-2000), four R&D programmes were implemented under BIOTEK under the coordinatorship of the BCCs. They were:

(I) Top-down research projects: National research projects directed by the National Scientific Research and Development Council for implementation by multi-institutions involved in biotechnology. A top-down project on rice genetic engineering for tungro virus resistance, for example, has led to the development of a transgenic rice that is resistant to one of the virus (RTSV) that is causing the disease. The project has also paved the way for other improvement of rice.

(II) Experimental developmental projects: Up-scaling projects that need to be developed after laboratory-scale R&D before they are commercialized. Public research institutions and universities are encouraged to forge partnerships with one or more companies in the private sector for undertaking these projects.

(III) Partnership programme: The programme involves partnership of a group of institutions in Malaysia with a prominent international center of excellence in relevant field of biotechnology. The Malaysia-MIT Biotechnology Partnership Programme is an example of a partnership programme that has entered its third year of implementation.

(IV) Bilateral programme: A supportive programme linking existing research interest/projects in Malaysia with a similar counterpart in another country. The support is mainly for exchange of information, training and transfer of technology.

Another programme that supports biotechnology R&D in Malaysia is the Intensified Research Priority Area (IRPA) under MOSTE. Although this programme is not under BIOTEK, advice and guidance of the Directorate are always required in implementing the programme on biotechnology. Issues that have been given priority in developing biotechnology projects are: Food security of the country; reducing the import bill on food and feed; and for wealth creation.

Among the projects that have been implemented are: genetic engineering of papaya for virus resistance and improved shelf life of fruits; gene transfer system for oil palm; genetic engineering of rubber; and genetic engineering of orchids for improved shelf life of flowers.

CONSUMER ACCEPTANCE, TRANSFER OF TECHNOLOGY AND GLOBALIZATION

As biotechnology is becoming more and more a commercial reality, its impact on consumers as well as producers is well recognized. The subject of GMOs is relatively new to Malaysian consumers. In general, the public is not well informed in the subject matter. As past surveys have shown, there is significantly more support among those who said they had heard a lot about GMOs than those who heard little about them.

The National Biotechnology Directorate is stepping up its effort to implement public awareness programmes on biotechnology during the 8 th Malaysia Plan (2001-2005). The programmes include arranging lectures at both a nationwide public forums as well as in schools; preparing and distributing phamplets about biotechnology; and promoting understanding on biotechnology through mass media.

With globalization coming our way, two new issues have to be addressed:

> *transfer of technology and the threat of globalization. Bulk of biotechnological research is done in the private sector, the relevant information and data are not publicly available, and most of the relevant technologies and materials are not in the public domain. With such a situation, a developing nation like Malaysia viewed*

technology transfer as a critical agenda since the nation is still far behind in expertise, infrastructures, and financial support. Developed nations must facilitate technology transfer to developing counterparts in order to face challenges thrown up by globalization. It is also the challenge to a country like Malaysia to acquire expertise and capability in biotechnology as it moves forward in the era of globalization.

FUTURE STRATEGIES

Two strategies have been adopted for future development of biotechnology in Malaysia. First relates to strengthening the existing programmes/projects by focusing on key research fields and key platform technologies. BIOTEK will continue to support capacity building such as infrastructure, human resource development (HRD) and management. The role of BCC programme will be enhanced through grouping for common technological needs under three main R&D areas, viz. agro-forestry, healthcare and energy/ environmental management. Technical competencies will be developed under a separate programme for technological development on in three identified areas, viz. genomics, bioprocessing and DNA diagnostics. The second strategy relates to developing new innovative projects which can ensure that research is translated into commercial products for global market place, thus making Malaysia more competitive in key biotechnology opportunity area. This will contribute to the national economy programme. This strategy is implemented through the establishment of the National Institute of Biotechnology Development Programme.

CONCLUSION

As the world moves into the era of globalization, where economy becomes borderless, enhanced efficiency, competitiveness, productivity and quality will take precedence in determining our success in the global arena. Malaysia is positioning herself to face the in-coming challenges being brought by globalization. Technology, such as biotechnology, will be a strong driving force energizing the shift into a knowledge-based economy. Intensive application of high technology and competency building will enhance this transformation process to position the country where it can look forward to emerge as a stronger economy. The country has no doubt about the benefits emanating from biotechnology, which was claimed by many as the technology of the 21 st century for national development. It has been recognized as a major growth industry worldwide.

Advances in biotechnology have spearheaded novel growth and development opportunities in a diverse range of industries such as pharmaceuticals, food processing, plant and animal production, and environmental management. The strategic plan for biotechnology developed by BIOTEK for the future should lead to a more dynamic and competitive biotechnology industry in Malaysia, capable of facing challenges of the next millennium and helping to contribute towards development of nation and move forward to realize the goal envisaged in achieving our Vision 2020.

9

Reducing Poverty through Agriculture

INTRODUCTION

At the dawn of the twenty-first century, about 900 million or 68 percent of the world's poor people live in Asia; about 500 million in South Asia, 300 million in East Asia, and 100 million in Southeast Asia and the Pacific (World Bank 1999). In addition, about 526 million people, including 160 million children, are undernourished2 (FAO 1999c). Not only do they lack access to sufficient money to buy food and other essentials, but neither do they have access to sufficient schooling, adequate housing, nor medical care. Those in rural areas are often short of water and fuel. Fertile land and water for farming are increasingly scarce. Urban poor lack money to buy enough food. That which they can afford may be deficient in protein and essential vitamins and minerals.

Although the absolute numbers of people living in poverty in Asia today are unacceptable, the situation could be much worse. In 1970, 60 percent of all Asians lived in poverty. That figure has been cut by almost half, with about one third of all Asians living in poverty in 2000. Also, countries such as Bangladesh, the People's Republic of China (PRC), and India have moved from periodic famines to almost self-sufficiency in food production. However, further efforts are needed to reduce poverty by another 50 percent by 2015, as targeted by world leaders during the World Food Summit in 1997.

The latter half of the twentieth century saw impressive advances in science and technology. We now have the capacity to apply this knowledge to reduce poverty and improve food security. This Working Paper discusses how biotechnology can be used to safely and effectively reduce poverty and improve food security in Asia.

Past Successes in Reducing Poverty Through Agricultural Science

Science and technology underpinned the economic and social gains in Asia over the past 30 years. In agriculture, these gains came to be known as the Green Revolution. Between 1970 and 1995, cereal production in Asia doubled, calorie availability per person increased by 24 percent, and real food prices halved (IFPRI 1997). Although the region's population grew by 1 billion people, overall food production more than kept pace with population growth (McCalla 1998). These food production increases were achieved largely by the cultivation of high-yielding varieties (HYVs) of rice and wheat, accompanied by expansion of irrigated areas, increases in fertilizer and pesticide use, and greater availability of credit.

The scientific basis for the Green Revolution stemmed from national and international research programs that led to the development and distribution of new HYVs, particularly of rice and wheat. The first generation of these new varieties were based on the introduction of new genes for dwarfing that made the HYVs shorter, more responsive to fertilizers, and less prone to falling over or lodging when fertilized and irrigated. Subsequent varieties also carried genes that gave increased pest and disease resistance and improved taste and grain quality.

The key elements in improving food security in Asia from 1970-95 were government policies reflecting a belief that investments in increasing agricultural productivity were a prerequisite to economic development. These national policies were supported by political leaders in Asia and by both the public and private sectors of the international community. This mix of supportive public policies, scientific discoveries, and public and private investments in rural Asia, particularly in irrigation, credit, and inputs, led to substantial reductions in poverty and improved food security throughout Asia over the past 30 years. Increased agricultural productivity, rapid industrial growth, and expansion of the nonfarm rural economy have all contributed to almost a tripling of per capita gross domestic product across Asia since 1970 (ADB 2000b, Pinstrup-Andersen and Cohen 2000).

Present Problems

Despite these successes, problems remain. The intensification of agriculture and the reliance on irrigation and chemical inputs has led to environmental degradation, increased salinity, and pesticide misuse. Deforestation, overgrazing, and overfishing also threaten the sustainable use of natural resources.

Green Revolution technologies had little impact on the millions of smallholders living in rainfed and marginal areas, where poverty is concentrated. Furthermore, the Green Revolution has already run its course in much of Asia. Wheat and rice yields in the major growing areas of Asia have been stagnant or declining for the past decade, while population continues to increase (Pingali et al. 1997). The key lessons learned from the Green Revolution are: (i) it has benefited farmers in irrigated areas much more than farmers in rainfed areas thus worsening the income disparity between the two groups, (ii) it overlooked the rights of women to also benefit from the technological advances, and (iii) it promoted an excessive of use of pesticides that are harmful to the environment. As countries became self sufficient in food, government investments declined in the agricultural sector and in science and technology across the region. This reflects a worldwide trend toward declining public investments in the rural sector and in agricultural research and development (R&D), nationally and internationally.

In Asia, private sector investments in the rural sector and related R&D have concentrated on export commodities. The downward trends in public investments by governments and development agencies in smallholder agriculture over the past decade have not been matched by a concomitant rise in private investments. Similarly, there is little (and few incentives for) private R&D on the food crops, livestock, fisheries, and aquaculture systems important for food security and poverty reduction in rural Asia.

Future Challenges

The population of Asia is projected to increase from 3.0 billion to 4.5 billion in the next 25 years. During the same period, the urban population will nearly double from 1.2 billion to 2.0 billion, as rural people move to the cities in search of employment. These increases will place massive pressure on developing member countries (DMCs) of ADB to increase food production.

Food demand is influenced by population growth, urbanization, income, and associated changes in dietary preferences. Urbanization and income growth frequently lead to shifts from a diet based on root crops (cassava, yam, and sweetpotato), sorghum, millets, and maize to rice and wheat, which require less preparation time, and to more meat, milk, fruits, vegetables, and processed foods. This dietary transition has already happened in much of the region (ADB 2000b). Meeting the food needs of Asia's growing and increasingly urbanized population requires increases in agricultural productivity and matching these increases to dietary changes and rising incomes.

To meet this demand, cereal production will need to be increased by at least 40 percent from the present level of about 650 million tons annually, most of which will have to come from yield increases. In addition, meat demand will double during the period (Pinstrup-Andersen et al. 1999). Production increases will have to be achieved by increasing yields in a sustainable way to conserve diminishing and degraded natural resources. Nearly all of these production increases will need to take place in DMCs themselves because on average 90 percent of the world's food is consumed in the country where it is produced. Food imports are not only expensive but discourage the creation of employment, which is badly needed in the rural areas.

In this millennium, we face a food, feed, and fiber production challenge in highly complex farming systems for several reasons:

(i) Water will become the most important limiting factor in agricultural production because the quality and quantity of water will decline as a result of pollution, forest degradation, and increased agricultural, domestic, and industrial use (ADB 2001).

(ii) Urbanization will mean the loss of agricultural land to residential and industrial development, and a decline in the number of farm workers.

(iii) Most farmers are poor with small landholdings.

(iv) Farming systems are commonly heterogeneous with mixes of food crops, livestock, and trees.

(v) About 70 percent of the cultivated land is rainfed with unreliable distribution and intensity of rainfall.

Thus, the increase in food production during the next 25 years will have to be achieved using less labor, water, and cultivated land. This can be done only if scientists can develop new crop varieties with high yield potential and high water use efficiency. New understanding of plant and animal genes may offer ways to increase crop yields to the levels required to adequately and sustainably feed the growing population in Asia.

Thus, developments in modern biotechnology could make extremely important contributions to future agricultural growth, food security, and poverty reduction. Increasing smallholder agriculture productivity will not only increase food supplies, but will reduce poverty and malnutrition, increase food access, and, improve living standards of the poor (McCalla and Brown 2000).

Rationale for ADB's Involvement in Biotechnology

Biotechnology has the potential to (i) increase crop and animal productivity; (ii) improve nutritional quality; (iii) broaden tolerance of crops for drought, salinity, and other abiotic stresses; and (iv) increase resistance of crops to pests and diseases.

These potential benefits will have significant impact in increasing food production and reducing poverty in DMCs if they can be applied to problems of the poor farmers in the tropics.

As with any technology, biotechnology brings with it potential risks. To maximize the benefits and minimize the risks, the introduction and use of biotechnology in DMCs or elsewhere must be thoughtfully managed by the public and private sectors.

The key risks that relate to the application of new developments in biotechnology for the public good are food and environmental safety, economic concentration, and intellectual property (IP) management. To explore the opportunities and risks of biotechnology, ADB has decided to undertake a comprehensive study on the use of biotechnology to reduce poverty and achieve food security in cooperation with the Australian Centre for International Agricultural Research (ACIAR)/ Australian Agency for Agricultural Development (AusAID).

There are many reasons to support this initiative:

(i) As a premiere development institution, ADB should be well informed on the development of biotechnology and the issues surrounding modern biotechnology.

(ii) Biotechnology has been identified by the Rural Asia Study as one of the emerging challenges in Asia (ADB 2000b).

(iii) Biotechnology has been identified by the Consultative Group on International Agricultural Research (CGIAR) as a powerful tool to fight poverty when used in conjunction with other agricultural research (Persley and Lantin 2000).

(iv) ADB is responsible for assisting the DMCs in dealing with potential risks of biotechnology and providing information on various biotechnology issues.

(v) ADB's Policy on Agriculture and Natural Resources Research recommends that ADB support research to increase farm productivity through modern biotechnology approaches, while giving appropriate consideration to indigenous and traditional knowledge (ADB 1995).

ADB's Poverty Reduction Strategy

While the need for further intensification of agricultural production in Asia is clear, intensification strategies must change to avoid adverse environmental impact and to reverse the effects of earlier practices. Enhanced, but inefficient use of irrigation and mineral fertilizers over the past three decades has had negative side effects such as soil salinity and nutrient leaching. With crop intensification, incidences of pests and diseases also increased (Pinstrup-Andersen and Cohen 2000). The following strategies are needed to meet the food demand in Asia over the next 25 years.

(i) Sustainable productivity increases in food, feed, and fiber crops.

(ii) Reducing chemical inputs of fertilizers and pesticides and replacing these with biologically based products.

(iii) Integrating soil, water, and nutrient management.

(iv) Improving the nutrition and productivity of livestock and controlling livestock diseases.

(v) Sustainable increases in livestock, fisheries, and aquaculture production.

(vi) Increasing trade and competitiveness in global markets.

As rural poverty persists in Asia, agriculture will play a prominent role in achieving equitable and sustainable rural growth in the twenty-first century. Even when rural people do not work directly in agriculture, they rely on nonfarm employment and income closely related to agriculture. Where there are large numbers of rural poor, agricultural growth is a catalyst for broad-based economic growth and development.

Agriculture's linkages to the nonfarm economy generate employment, income, and growth in the rest of the economy. A healthy agricultural economy also offers incentives for natural resource conservation (Pinstrup-Andersen and Cohen 2000).

The ADB Poverty Reduction Strategy (ADB 1999) sees three factors, pro-poor, sustainable economic growth; good governance; and social development as key elements for reducing poverty. Biotechnology may contribute toward achieving the poverty reduction goals in several components of the strategy:

(i) Increasing priority for agriculture and rural development.

(ii) Addressing environmental considerations.

(iii) Increasing the public and private sector roles in poverty reduction.

(iv) Encouraging regional and subregional cooperation.

(v) Coordinating efforts of funding agencies.

The challenge is how to use new scientific developments such as biotechnology, together with information and communications technology, to make the complex agricultural systems of Asia more productive and sustainable. Good governance is also crucial to ensure that new agricultural biotechnology reaches the poor.

Key Questions

It is in the context of this complex and evolving situation, that this report will address three key issues:

(i) What are the potential benefits and risks of agricultural biotechnology on human health, the environment, and agriculture; how can we minimize the risks and enhance the benefits?

(ii) How can we use biotechnology to reduce poverty and achieve food security in Asia?

(iii) What policies and strategies should be considered by funding agencies, including ADB, to support biotechnology for the benefits of small farmers in Asia?

SCIENTIFIC CONTEXT

Promethean Science

The pace of change in modern science has led some to term it Promethean science, acknowledging both its risks and benefits (Serageldin and Persley 2000). Modern science encompasses new developments in the biological, physical, and social sciences. In biology, discoveries over the past 20 years allow the better understanding of the structure and function of human, animal, and plant genes.

At the same time, new discoveries in the physical sciences underpin the revolution in information and communications technologies. Geographic information systems enable characterization of agro-ecosystems and offer means by which new technologies can be customized to the needs of particular agro-ecosystems. The biological and physical sciences also interact in new ways. For example, the ability to analyze large volumes of data is a critical component of various genome projects that are mapping all the genes in an organism, as in the Human Genome Project.

New developments in the social sciences underpin community participation in technology development and evaluation (sometimes termed agro-ecological methods). Participatory methods developed in the social sciences can help in understanding problems and the researchable issues, particularly those of small farmers operating in marginal environments. They may also be used to clarify the concerns of rural and urban dwellers in regard to the deployment of new technologies, including the products of biotechnology.

Integration of all branches of modern science and traditional knowledge is required to develop knowledge-intensive solutions to the problems of rural Asia. These solutions need not be only technically feasible but also socially acceptable. Indeed, the potential value of modern science to agriculture and the environment in Asia will require the efforts of all stakeholders, including civil society, farmer cooperatives, producers, consumers, governments, and development agencies.

Definition, History, and Scope of Biotechnology

Biotechnology, broadly defined, includes any technique that uses living organisms, or parts of such organisms, to make or modify products, to improve plants or animals, or to develop microorganisms for specific use.

It ranges from traditional biotechnology to the most advanced modern biotechnology. Biotechnology is not a separate science but rather a mix of disciplines (genetics, molecular biology, biochemistry, embryology, and cell biology) converted into productive processes by linking them with such practical disciplines as chemical engineering, information technology, and robotics.

Modern biotechnology should be seen as an integration of new techniques with the well-established approaches of traditional biotechnology such as plant and animal breeding, food production, fermentation products and processes, and production of pharmaceuticals and fertilizers (Doyle and Persley 1996).

The key components of modern biotechnology are listed below.

(i) *Genomics*: The molecular characterization of all genes in a species.

(ii) *Bioinformatics*: The assembly of data from genomic analysis into accessible forms, involving the application of information technology to analyze and manage large data sets resulting from gene sequencing or related techniques.

(iii) *Transformation*: The introduction of one or more genes conferring potentially useful traits into plants, livestock, fish and tree species.

(iv) Genetically improved organism.

(v) Genetically modified organism (GMO).

(vi) Living modified organism (LMO).

(vii) *Molecular breeding*: Identification and evaluation of useful traits in breeding programs by the use of marker-assisted selection (MAS);

(viii) *Diagnostics*: The use of molecular characterization to provide more accurate and quicker identification of pathogens; and

(ix) *Vaccine technology*: The use of modern immunology to develop recombinant deoxyribonucleic acid (rDNA) vaccines for improved control of livestock and fish diseases (Doyle and Persley 1999).

Biotechnology consists of a gradient of technologies, ranging from the long-established and widely used techniques of traditional biotechnology to novel and continuously evolving modem biotechnology techniques. During the 1970s scientists developed new methods for precise recombination of portions of deoxyribonucleic acid (DNA), the biochemical material in all living cells that governs inherited characteristics, and for transferring portions of DNA from one organism to another. This set of enabling techniques is referred to as rDNA technology or genetic engineering.

Modern biotechnology presently includes the various uses of new techniques of rDNA technology, monoclonal and polyclonal antibodies, and new cell and tissue culture methods. A chronology of the development of modern biotechnology is given in Table 2.1. Over the past two decades the number of significant advances in modern biotechnology for understanding and modifying the genetics of living organisms has increased dramatically. That has led to greatly increased interest and investment in biotechnology, and increasing concerns as to the power of the new technologies and their safety.

Economic Concentration in Agricultural Biotechnology

Modern biotechnology R&D has been conducted in an institutional and economic environment that differs significantly from the development of the earlier Green Revolution technologies. While the latter were essentially the prerogative of public research institutions and philanthropic foundations, the application of modern biotechnology to agriculture is essentially a competitive, commercial endeavor in which powerful private sector interests compete. Similarly, while the Green Revolution technologies were essentially dedicated to the public, the strengthening and extension of IP protection, particularly since the conclusion of the Uruguay Round of trade negotiations has increased the private character of biotechnologies.

Multinational companies in the seed, agricultural chemical, pharmaceutical, and food-processing industries play a major role in biotechnology research. They have invested heavily in in-house research facilities, commissioned research, taken equity positions in new biotechnology firms, and entered into contractual arrangements with public research institutions or universities. As a result of mergers and acquisitions in the past few years, the development of new biotechnology applications in agriculture has become increasingly concentrated in the hands of a decreasing number of companies. The dominant companies that operate within global markets are Aventis, AgrEvo, Dow, DuPont, Monsanto, and Syngenta.

Table : The Evolution of the Science of Genetics, Leading to Modern Biotechnology

1866 Mendel postulates a set of rules to explain the inheritance of biological characteristics in living organisms.

1900 Mendelian law rediscovered after independent experimental evidence confirms Mendel's basic principles.

1903 Sutton postulates that genes are located on chromosomes.

1910 Morgan's experiments prove genes are located on chromosomes.

1911 Johannsen devises the term "gene", and distinguishes genotypes (determined by genetic composition) and phenotypes (influenced by environment).

1922 Morgan and colleagues develop gene mapping techniques and prepare gene map of fruit fly chromosomes, ultimately containing over 2000 genes.

1944 Avery, MacLeod and McCarty demonstrated that genes are composed of DNA rather than protein.

1952 Hershey and Chase confirm role of DNA as the basic genetic material.

1953 Watson and Crick discover the double-helix structure of DNA.

1960 Genetic code deciphered.

1971 Cohen and Boyer develop initial techniques for rDNA technology, to allow transfer of genetic material from one organism to another.

1973 First gene (for insulin production) cloned, using rDNA technology.

1974 First expression in bacteria of a gene cloned from a different species.

1976 First new biotechnology firm established to exploit rDNA technology (Genentech in USA).

1980 USA Supreme Court rules that microorganisms can be patented under existing law (Diamond v. Chakrabarty).

1982 First rDNA animal vaccine approved for sale in Europe (colibacillosis).

First rDNA pharmaceutical (insulin) approved for sale in USA and UK.

First successful transfer of a gene from one animal species to another (a transgenic mouse carrying the gene for rat growth hormone).

First transgenic plant produced, using an agrobacterium transformation system.

1983 First successful transfer of a plant gene from one species to another.

1985 US Patent Office extends patent protection to genetically engineered plants.

1986 Transgenic pigs produced carrying the gene for human growth hormone.

1987 First field trials in USA of transgenic plants (tomatoes with a gene for insect resistance).

First field trials in USA of genetically engineered microorganism.

1988 US Patent Office extends patent protection to genetically engineered animals.

First GMO approved. Human genome mapping project initiated.

1989 Plant genome mapping projects (for cereals and *Arabidopsis*) initiated.

2000 Plant genome mapping projects for rice and Arabidopsis completed, and about 44 million hectares of land planted to GMO crops.

DNA = deoxyribonucleic acid, GMO = genetically modified organism, rDNA = recombinant

DNA, UK = United Kingdom, USA = United States of America.

Source: Adapted from Persley (1990).

Biotechnology R&D has been concentrated in a limited number of industrialized countries, with the United States (US) in the lead in financial and human resources. A growing number of developing countries have invested in biotechnology R&D, but the amounts are small compared to the sums invested by private companies in the industrial world. While private sector investment in agricultural research in general is increasing in developed countries, there is still little private sector biotechnology research effort in developing countries, particularly in Asia (Pinstrup-Andersen and Cohen 2000).

The commercialization and distribution of new agricultural biotechnology products, particularly transgenic crops, is also concentrated in Organisation for Economic Co-operation and Development (OECD) member countries, with a few exceptions (James 2000). These products are for the most part crops of economic importance in industrial country agricultural production and in world trade in agricultural commodities, mainly soybean, maize, cotton, and canola. During the past five years, the area under GMOs3 has increased rapidly from 1.7 million ha to 44.4 million ha, 75 percent of which are in the US. The remaining 25 percent are distributed in both developed and developing countries, including Argentina, Australia, Bulgaria, Canada, PRC, France, Germany, Mexico, Portugal, Romania, South Africa, Spain, Ukraine, and Uruguay (Appendix 2).

Relatively little biotechnology research is being undertaken on the problems of small farmers in rainfed and marginal lands. Neither is there much interest in Asia's basic food crops: rice, tropical maize, wheat, sorghum, millet, banana, cassava, groundnut, oilseed, potato, sweetpotato, and soybean. These are considered *orphan crops* because of the private sector's reluctance to work on them. That focus is unlikely to change because of the perception that investments in such orphan crops and from working on problems of small farmers yield limited returns. To participate more fully in the biotechnology revolution, Asian governments will need to expand their capacities to undertake biotechnology research linked to the problems of small farmers and orphan crops. In certain situations, however, there may be opportunities to purchase, license, or import technology applicable in Asia.

Applications of Biotechnology

The applications developed from the new methods in biotechnology place them within the continuum of techniques used throughout human history in industry, agriculture, and food processing. Thus, while modern biotechnology provides powerful new tools, they are used to generate products that fill similar roles to those produced with more traditional methods.

There is now increasing use of modern molecular genetics for genetic mapping and MAS as aids to more precise and rapid development of new strains of improved crops, livestock, fish, and trees. Other biotechnology applications such as tissue culture and micropropagation are being used for the rapid multiplication of horticultural crops and trees. New diagnostics and vaccines are being widely adopted for the diagnosis, prevention, and control of fish and livestock diseases.

The science of genomics (the molecular characterization of all the genes in a species) has dramatically increased knowledge of plant genes and their functions. The new technologies enable greatly increased efficiency of selection for useful genes, based on knowledge of the biology of the organism and the role of specific genes in regulating particular traits. This will enable more precise selection of improved strains.

These techniques may be used for more efficient selection in conventional breeding programs. They may also be used for the identification of genes suitable for use in the development of transgenic crops. Thus far, scientists have completed genomic study on rice through the cooperative efforts of several international and private sector institutions led by Japan.

Modem biotechnology permits increased precision in the use of new techniques and a shorter time to produce results. For example, plant breeders and molecular biologists can collaborate to transfer to a highly developed crop variety one or two specific genes to impart a new character such as a specific kind of pest resistance.

New techniques of modern biotechnology accelerate plant and animal breeding. They offer possible solutions to previously intractable problems and difficult targets such as drought tolerance, and enable the development of new products. These products may include more nutritious food, crop varieties with improved tolerance for pests and diseases, and animal vaccines.

It is important to provide appropriate regulatory mechanisms to ensure that products produced by modern biotechnology are as safe as the products of traditional biotechnology. That is especially so when the products are GMOs that might interact with the environment.

Our knowledge of genes is not matched by our knowledge of the gene-environment interaction or potential impacts of biotechnology on the environment. However, many of the Green Revolution technologies were also introduced without such understanding. At present, there is widespread distrust of biotechnology and the public needs to be engaged in dialogue before the technology is disseminated widely.

PRESENT STATUS OF AGRICULTURAL BIOTECHNOLOGY IN ASIA

Selected Country Experiences

Several governments in Asia are committed to the use of modern biotechnology in agriculture. They have devoted significant human and financial resources to this policy over the past two decades. Some illustrations of current activities in selected countries are given below.

The PRC accords high priority to biotechnology to increase food production and improve product quality in an environmentally sustainable manner. The PRC has moved quickly to adopt new biotechnologies, particularly genetically modified crops. The country is rapidly increasing its expenditure on biotechnology R&D. Over 103 genes have been evaluated for improving traits in 47 plant species. New traits have been introduced and evaluated in field tests on rice, wheat, maize, cotton, tomato, pepper, potato, cucumber, papaya, and tobacco. A variety of traits have been targeted. They include resistance to diseases, pests, and herbicides, and quality improvement. Approximately 50 genetically modified varieties have been approved for environmental release, or small-scale field testing in the PRC. A few new genetically improved varieties have been approved for largescale commercial production. The most widespread are new pest-resistant varieties of cotton that are being widely cultivated by farmers. These were grown commercially by approximately 3 million farmers on approximately 500,000 ha in 2000. Several new products are in the pipeline for potential commercialization (Zhang 2000).

India has allocated large public resources toward human resources development and infrastructure in biotechnology. In the early 1980s, the Government of India created a Department of Biotechnology to promote the use of new biotechnologies in industry, medicine, and agriculture. Current R&D efforts in India are directed toward increasing agricultural productivity, bioremediation in the environment, medical and industrial biotechnology, and bioinformatics (Sharma 2000). R&D priorities in agriculture include new regeneration techniques for the rapid multiplication of citrus, coffee, mangrove, vanilla, and cardamom.

Cardamom yield has increased 40 percent through the use of tissue culture.

There is substantial private sector participation in biotechnology in India, for example, in the seed sector, the veterinary products sector, and bioinformatics linked with the booming information technology sector (Dhawan 2001). Further, organizations like the M.S. Swaminathan Research Foundation have developed innovative approaches such as the ADB-supported BioVillages program that is fostering the growth of new income-raising technologies. Emphasis is on developing small-scale bioindustries for women (Lakshmi 2001).

Indonesia has placed a high priority on biotechnology over the past 15 years. The Government has designated three National Biotechnology Centers to coordinate R&D in agriculture, medicine, and industrial microbiology. Applications of biotechnology to agriculture are primarily the responsibility of the Agency for Agricultural Research and Development (AARD). A National Committee on Biotechnology advises the minister in developing guidelines for government policy in the promotion of biotechnology.

In recent years there has been an extensive training program within Indonesia and abroad to upgrade skills of scientists involved in biotechnological research. In the 1980s, a major World Bank loan of over $100 million financed the creation of three inter-university centers for agricultural, medical, and industrial biotechnology. More recently, a current World Bank loan is financing facilities for agricultural biotechnology within AARD.

Crop improvement efforts using modern biotechnology started in Pakistan in 1985, when a training course was held on recombinant DNA. Work is now concentrated on chickpea, rice, and cotton. Field evaluation is hampered by lack of biosafety regulations. There is some private investment in R&D of agricultural biotechnology. The government controls testing, multiplication, distribution, and biosafety issues for genetically modified crops. Pakistan lacks firm policy and regulations regarding intellectual property rights (IPR) and patents involving biotechnology, and biosafety regulations for GMOs (Zafar 2001).

The Philippines began its modern biotechnology programs in 1980 with the creation of the National Institutes of Molecular Biology and Biotechnology in Los Baños, with a focus on agricultural biotechnology. In 1997, the Agriculture Fisheries Modernization Act recognized biotechnology as a major strategy to increase agricultural productivity. The Act provided a budget for agricultural biotechnology of almost $20 million Present Status of Agricultural Biotechnology in Asia 21 annually for the next 7 years (4 percent of the total R&D budget). In 1998, the government funded these five high-level biotechnology research projects to develop

(i) new varieties of banana resistant to banana bunchy top virus and papaya resistant to ringspot virus,

(ii) delayed ripening papaya and mango,

(iii) insect-resistant maize,

(iv) marker-assisted breeding in coconut, and

(v) coconut oil with high lauric acid content.

Nongovernment organizations (NGOs) and other groups concerned about the safety of GMOs have been vocal in the Philippines. This is affecting field-testing and commercialization of transgenic crops. Products in the regulatory pipeline include new varieties of insect-resistant maize and insect-and disease-resistant rice.

Thailand is focusing on the applications of biotechnology to traditional foods, fruits, and export commodities such as shrimp. R&D priorities are to increase production and reduce production cost on crops such as rice, cassava, sugarcane, rubber, durian, and orchids. An early success in Thailand has been the development of new molecular diagnostics for the diagnosis and control of virus diseases in shrimp. These diseases cost the shrimp export industry over $500 million in lost production in 1996. The development and commercial use of the new diagnostics prevents the loss of an estimated 20–50 percent of annual production, a saving of at least $100 million per year.

There are also active agricultural biotechnology programs in Bangladesh, Malaysia, Nepal, Sri Lanka, Singapore, and Viet Nam.

Regional and International Programs

International Agricultural Research Centers

International R&D programs using modern biotechnology are being conducted by the international agricultural research centers (IARCs), particularly the International Rice Research Institute (IRRI), International Maize and Wheat Improvement Center (CIMMYT), International Crops Research Institute for the Semi-Arid Tropics (ICRISAT), International Livestock Research Institute (ILRI), and International Service for National Agricultural Research (ISNAR). The Center for International Forestry Research also uses biotechnology in the characterization of forest diversity in its Asian program. The International Center for Living Aquatic Resources Management is using new technologies in the improvement of fisheries and aquaculture systems. ILRI is initiating a program on Asian livestock improvement. The CGIAR centers invest approximately $30 million per year in modern biotechnology.

ISNAR Biotechnology Service

The ISNAR Biotechnology Service (IBS), with Japanese support, has been assisting selected Asian countries in developing human resources for managing biotechnology research programs or institutions. IBS has developed specialized courses to enhance the capacity and competency of managers, focusing on strategy building, priority setting, managing biosafety and regulatory aspects, resource generation and deployment, product delivery, and information sharing as well as the establishment and management of linkages.

International Service for the Acquisition of Agri-biotech Applications

The International Service for the Acquisition of Agri-biotech Applications (ISAAA) is brokering public-private sector partnerships to facilitate technology transfer. It has current regional projects to increase the productivity of maize, papaya, and sweetpotato. ISAAA is also establishing a new Asian knowledge center for crop biotechnology, based in the Philippines, in partnership with CAB International and the Southeast Asian Regional Center for Graduate Study and Research in Agriculture. Its purpose is to make available timely and balanced information on the risks and benefits of crop biotechnology to interested parties in Asia. In doing so, it aims to provide training and study tours highlighting experiences not only in research but also with biosafety and intellectual property issues. Present Status of Agricultural Biotechnology in Asia 23

Activities of Funding Agencies

The major external sources of assistance for agricultural biotechnology in the Asia/Pacific region are ADB, Australia, the Rockefeller Foundation, the United Nations, the United States Agency for International Development (USAID), and the World Bank.

Asian Development Bank

ADB has made several strategic and innovative investments in agricultural biotechnology over the past decade. These investments have been in the form of loans and technical assistance projects.

Components of Agriculture and Science and Technology Loan Projects

Several governments have requested ADB financial support for human resource development, laboratory facilities, and equipment for agricultural biotechnology programs. These programs have been integrating new applications of biotechnology into their conventional agricultural R&D programs. The applications include the use of new molecular diagnostics for pests and diseases and MAS for crop and livestock breeding. Such components are being supported under ongoing ADB projects in the Philippines, Sri Lanka, and Thailand. ADB has provided other loans and grants in closely related areas such as in integrated pest management in cotton under the Cotton Development Project in Pakistan.

ADB Regional Technical Assistance Projects

ADB has provided regional technical assistance grants for the development of the three regional crop biotechnology networks over the past decade. The networks are the Asian Rice Biotechnology Network (ARBN), initiated in 1993, the Asian Maize

Biotechnology Network (AMBIONET), initiated in 1998 and the recently established Asian Semi-Arid Crops Network.

These networks have been influential in developing capacity to use new techniques in crop breeding for the major cereal crops (rice, maize) and the crops important in the semi-arid regions (sorghum, pigeonpea, and groundnut). The networks are managed by three IARCs (IRRI, CIMMYT, and ICRISAT) that work with national re search institutes to implement the programs on rice, maize, and semi-arid crops.

A recent evaluation of ADB's research investments through the international centers reported favorably on the achievements of the rice and maize biotechnology networks, and recommended further support for this type of research cooperation.

The networks provide a platform by which countries and the IARCs can collaborate in the use of the new tools of functional genomics to identify genes that control important traits such as drought and salinity tolerance. This will speed up the breeding of crop varieties with these characteristics that have been difficult to address through conventional breeding. There will be increasing opportunities for cooperation between the networks, as more knowledge is gained on the commonality of genes between species and their control.

Asian Rice Biotechnology Network

The rice growing countries in Asia recognized in the early 1990s that biotechnology could provide powerful new tools for rice improvement. These tools included tissue culture, wide hybridization, molecular markers, and genetic engineering. The ARBN was initiated in 1993 to provide a vehicle for collaborative research in these areas with universities and national rice research institutes.

The objectives of ARBN are to (i) promote human resource and infrastructure development for biotechnology at selected national agricultural research systems (NARSs) institutes through joint research and training coordinated by IRRI, and (ii) generate biotechnology tools and products for use by NARSs through IRRI research and infrastructure development. ARBN was established with funds provided by ADB and the German Government's Budenministerium fur Technische Zussamenarbeit, which has been supporting IRRI's development of biotechnological products of direct use to NARSs. ADB supports research and infrastructure development at NARSs with additional funding for training and shuttle research by NARSs scientists at IRRI.

Australian Government

Over the last 15 years, ACIAR has entered into more than 100 biotechnology R&D partnerships with at least 10 Asian countries in support of more than 600 active or completed projects. The emphasis of this work has been on developing diagnostics and vaccines for a large suite of diseases of tropical livestock, with some recent work on fish and shrimp being undertaken. Most of these projects have been implemented through government programs, but NGOs are becoming increasingly involved.

Molecular marker methods for identifying disease-resistant genes and prolificacy in livestock have also been developed. Several projects on biotechnology for rumen manipulation have been carried out.

Asian Maize Biotechnology Network

Member countries of AMBIONET are PRC, India, Indonesia, Philippines, and Thailand. The network was established in 1998 with financial support from the ADB. AMBIONET emphasizes developing molecular markers for specific traits and using them in the selection of new maize varieties. The goals of AMBIONET are to:

- *increase the scientific capacity of Asian maize and biotechnology programs to ensure higher, more stable, and more sustainable maize productivity for farmers in Asia, and thereby help meet the region's rapidly growing demand for maize;*
- *develop sustainable, environmentally friendly, and natural resource conserving maize production systems;*
- *empower national programs to effectively use modern biotechnology for maize improvement;*

- *strengthen the ability of national programs to identify and overcome the key production constraints faced by maize farmers in the region; and*
- *generate and distribute improved maize cultivars, and implement improved crop management strategies, in collaboration with existing national program personnel, and by using facilities and other resources more effectively to avoid duplication; and ensuring the long-term sustainability of integrated maize and biotechnology research programs in participating countries.*

Crops and forestry work has focused on development of diagnostics for diseases (viral, fungal, mycoplasma, and bacterial) and contaminants in tropical crops, and the application of biofertilizers, bioremediation technology, and biofumigants. Molecular markers have been developed for the improvement of cereals and tree species. In cooperation with IRRI, attempts are being made to develop apomixis systems for rice. Tissue culture methods for micropropagation and conservation of several species, including sweetpotato, taro, tropical fruits, coconut, green tea, and tree species such as mangrove are being developed.

Eight of ACIAR's current or completed projects have included aspects of plant genetic engineering, with target crops being cereals and pulses, groundnut, and several tropical fruits. Target characters include virus resistance and quality defects related to ripening processes. These collaborative projects were developed at the request of the Asian countries, which fully approved regulatory procedures. ACIAR also provides core funding to many CGIAR and other IARCs, a proportion of which is applied to biotechnology R&D.

Australian support has been provided through AusAID for an Association of Southeast Asian Nations (ASEAN)/Australia regional biotechnology network, mainly concerned with food, microbial, and industrial biotechnology. AusAID also supports the characterization and conservation of genetic resources of taro and forest genetic resources in the South Pacific. AusAID has supported several biotechnology seminars. A major seminar in 1989 at the Australian Academy of Sciences on the potential of agricultural biotechnology in international development reported on the outcomes of a joint study cosponsored by AusAID, ACIAR, ISNAR, and the World Bank (Persley 1990a, 1990b).

In-the area of human resources, the Crawford Fund has sponsored several master classes in biotechnology for senior policymakers over the past decade.

The Rockefeller Foundation

Operational from 1984-1999, the Rice Biotechnology Network sponsored by the Rockefeller Foundation was very active and successful in the region. The program's two objectives were to (i) to create biotechnology applications to produce improved rice varieties suited to developing country needs, and (ii) to train scientists in developing countries to use the techniques and adopt them to their own objectives. A network of about 200 senior scientists and 300 trainee scientists from all the major rice producing countries of Asia and a number of industrialized countries participated. The program was funded at approximately $5 million per year for 15 years.

Another outcome of the program was the development of promising new technologies for the control of rice pests and diseases and improving the nutritional quality of rice. The work is being continued by national governments, ARBN, and IRRI to do the field evaluation and distribute new rice varieties in Asia.

The Rockefeller Foundation presently concentrates its biotechnology programs in Asia on drought tolerance in rice and maize. It is also examining innovative means of dealing with access to technologies and IP issues.

United Nations Agencies

The Food and Agriculture Organization (FAO) is giving high priority to biotechnology within its Asia/Pacific regional programs (FAO 2000). During 1989-1993, the United Nations Development Programme financed the establishment of biotechnology centers in eight countries (PRC, India, Indonesia, Republic of Korea, Malaysia, Pakistan, Philippines, and Thailand) to share rDNA techniques in animal improvement, embryo transfer, and disease control. In 1994 the United Nations Industrial Development Organization (UNIDO) established an International Center for

Genetic Engineering and Biotechnology in New Delhi, India, to assist Asian countries in the applications of biotechnology to important crops of the region.

United States Agency for International Development

The USAID is supporting applications of biotechnology through bilateral activities in several countries, including India, Indonesia, and Pakistan. USAID is also providing specific support for biotechnology applications within the programs of IARCs. This includes support for research by IRRI on improving the nutritional quality of rice by increasing its vitamin A and iron content, and support for research by the Tata Energy Research Institute on the development of *golden* mustard.

World Bank

The World Bank has supported the development of infrastructure and human resource development for biotechnology in several Asian countries over the past 15 years. This support has come through loans in the agricultural sector, science and technology, and education. There are currently substantial components for biotechnology within agricultural technology projects in India, Indonesia, and Pakistan. The Inter University Centers for Biotechnology in Indonesia were built with a $150 million loan in the 1980s. The Republic of Korea also used World Bank loans to develop its infrastructure in biotechnology. The World Bank is also one of the main financial supporters of CGIAR centers. The centers invest about 10 percent of their total annual budget of $340 million in the applications of biotechnology.

RISK ASSESSMENT AND MANAGEMENT

Public Perception of Risks

As with any science and technology, biotechnology can bring with it benefits and risks. It is the risks of agricultural biotechnology that have received widespread publicity in the media even though biotechnology has also been applied to health and industrial sectors. Environmental NGOs have been particularly vocal in taking issue with the new technologies derived from or incorporating GMOs. As a consequence, in the public debate biotechnology has become synonymous with GMOs, although they are only one of the many products of biotechnology. Curiously, biotechnology and GMOs in health-care products now in widespread use (insulin, hepatitis vaccine, medication for cardiovascular disease, etc.) or for industrial purposes such as bioremediation have elicited no such controversy. This can probably be attributed to the lack of benefits to consumers in the first generation of genetically modified (GM) crops. The main focus was on herbicide and insect resistance that benefited farmers, seed producers, and chemical companies. It is expected that the next generation of genetically modified foods will benefit consumers, nutritionally or from taste or storage benefits, and accordingly may be better accepted.

A number of food-related crises in recent years have made consumers particularly sensitive about food safety issues. Health and food safety concerns are again at the forefront in Europe following additional cases of mad cow disease *(bovine spongiform encephalopathy)* and the banning throughout the European Union of blood and bone meal in feed for all animals. These crises have not been caused by GMOs, but by the intensification of agriculture and food production, a fact that appears to have escaped public attention. In Europe in particular, demands have been made for informative food labeling so that consumers may, if they wish, avoid genetically modified foods.

The anti-GMO movement reveals profound mistrust of developments in science and technology and of the forces seen to be driving them. Genetically modified crops such as maize, sorghum, cotton, and canola have been widely grown for the last five years, yet no harmful effects on human health or the environment have been detected. That was one of the conclusions of the OECD-sponsored conference held in Edinburg, UK, in 2000. However, it is generally agreed that government and the private sector are responsible for monitoring the long-term effects of GMOs on human health and the environment.

Potential Risks of Biotechnology

The risks associated with modern biotechnology fall into four categories: food safety, environmental, socioeconomic, and ethical. Some of these concerns relate to potential risks inherent in modern biotechnology and can be described as technology-inherent

(Leisinger 2000). Others are related more to value systems or cultural practices and can be described as technology-transcending.

Food Safety Concerns

The potential risks of biotechnology on human health may include toxic reactions, increased cancer risks, food allergies, food contamination, and antibiotic resistance. There is also concern that GMOs in animal feed might present a health risk for consumers, or for the animal itself. Consumers are also concerned about the long-term health effects of genetically modified foods.

To address food safety concerns, the following safeguards have been adopted by some countries:

(i) Some countries have regulatory procedures, institutions, and infrastructure in place to ensure food safety (OECD 2000). These regulations cover all aspects of the food chain, from farm inputs (including animal feed, feed additives, pesticides, fertilizers, veterinary drugs) through production and processing (including agricultural products, processed food, novel foods, food additives), to transportation, storage, and distribution.

(ii) Formal science-based procedures for risk analysis of food have been adopted by some countries. These continue to evolve with new scientific information about food safety, emerging pathogens, new technology, and consumer demands for a high level of public health protection. Generally, the procedures conform to international standards set by the Codex Alimentarius Commission of FAO.

(iii) Some countries have adopted a variety of approaches to regulate genetically modified foods, either by applying existing food safety measures or enacting new legislation.

(iv) In response to consumer demands, a growing number of countries are introducing labeling. Opinions differ on whether labeling should be mandatory or voluntary, as well as on acceptable tolerance levels and on the type of information to be used. Two labeling issues currently being addressed are segregation and traceability. Segregation relates to the ability to attest to the separation of genetically modified and non-genetically modified crops (as in the case of soybean in processed foods). Traceability means being able to attest to the origins of food products. Traceability is of particular importance in the current controversy related to mad cow disease. Low cost, quick tests are now available to determine the presence of genetically modified products in food.

Environmental Concerns

A number of biotechnology applications are not seen to present any new threats to the environment. That is the case with tissue culture, diagnostics, and market-selected plant breeding. On the other hand, there is fear of potential risks from the release of GMOs into the environment. The potential risks of GMOs on the environment may include increased pesticide residues, genetic pollution, damage to beneficial insects, creation of superweeds and superpests, creation of new viruses and bacteria, and genetic bioinvasion.

To address environmental concerns, some countries have adopted the following safeguards:

(i) Most current biotechnology applications in tissue culture and micropropagation, diagnostics and vaccines, and markerassisted plant breeding are subject to existing regulations. They include phytosanitary regulations and plant quarantine, varietal certification of seeds, and veterinary product regulations. These regulations usually conform to international standards, guidelines, or recommendations such as those set by the International Plant Protection Convention or the International Bureau of Epizootics for animal products.

(ii) A number of countries have introduced new requirements and procedures for the environmental release of GMOs. Procedures for hazard identification and risk assessment of GMOs are now well-established in most OECD countries, and in some Asian countries.

Socioeconomic Concerns

Modern biotechnology R&D has been conducted in an institutional and economic environment that differs significantly from the development of the earlier Green Revolution technologies. While the latter were essentially the prerogative of public research institutions and philanthropic foundations, developments in biotechnology have been driven essentially as a competitive, commercial endeavor in which powerful private sector actors compete.

The major socioeconomic risk of agricultural biotechnology stems from the fact that the research, development, commercialization, and distribution of new biotechnological products have been carried out mainly in developed countries by a few, large, multinational companies.

These companies have focused on temperate crops for large farmers in developed countries. Undertaking R&D on Asia's basic food crops for small farmers in rainfed and marginal areas is of little interest because they see limited returns from such investments.

If this trend continues, modern biotechnology will aggravate the income disparity between developed and developing countries, and between large and small farmers. Increased public investments in agricultural biotechnology are necessary to ensure that small, poor farmers have access to biotechnology. Governments must also address the potential gender disparity in the access to technology, and the negative impact specific to women, since no technology is gender neutral. Lessons from earlier agricultural technological changes should be used as a caveat.

Ethical Concerns

Ethical issues may stem from the uneven or inequitable impact of an expanding global economy; from the national social, economic, and politiRisks Assessment and Management 41 cal context; or from individual values. Indeed, it may be difficult to reconcile personal values with what a majority regard as the common good. One of the ethical concerns raised by biotechnology—and particularly GMOs—is that it is unnatural and an unwarranted tampering with nature. However, seen in historical perspective, most technology developments in agriculture over the centuries have involved, in one way or another, efforts to overcome the vagaries of nature.

Risk Management

To address the potential risks of biotechnology in agriculture, there is a need for Asian countries to establish effective biosafety procedures for the development, testing, and release of the technologies.

Current and Emerging Regulatory Systems for Biotechnology

A growing number of international organizations have become involved in food and environmental safety. Concerted international efforts to develop agreed upon scientific approaches to biosafety date from 1975, when a group of scientists from the United States National Academy of Sciences expressed concern regarding the potential biological risks of rDNA molecules. Efforts to ensure the science-based, case-by-case risk identification, assessment, and management of GMOs have thus been pursued for some 25 years.

In the industrialized world, OECD has been instrumental in forging agreements on scientific principles regarding the safe applications of rDNA organisms in industry, agriculture, and the environment. Since the 1980s, OECD has established guidelines, first for (i) laboratory-based experimentation, then (ii) for small- and large-scale field trials of genetically modified plants and organisms, and finally (iii) for their commercialization and release into the environment or food chain.

During the 1990s, a growing number of international organizations including the United Nations Environmental Programme (UNEP), UNIDO, World Health Organization (WHO), and FAO have become involved in activities related to biosafety. Both UNIDO and UNEP have developed biosafety guidelines.

FAO and the WHO are jointly responsible for the Codex Alimentarius Commission, which sets international standards for food safety. Codex has set up a new Intergovernmental Task Force on Foods Derived from Biotechnology to develop general principles for risk analysis and to provide guidance on risk assessment.

Other international organizations conducting work on the safe use of biotechnology include the International Centre for Genetic Engineering and Biotechnology, the United Nations

Conference on Trade and Development, and, for many aspects of animal as well as human health, the International Bureau of Epizootics.

The Convention on Biological Diversity and the Cartagena Biosafety Protocol

Preoccupation with the protection of biological diversity, necessary for sustaining agriculture and food production and, indeed, life itself, became a topic of international debate during the 1980s. As a result, FAO established the International Undertaking on Plant Genetic Resources as a nonlegally-binding agreement for cooperation in the conservation of genetic material. The agreement was based on the universally accepted principle that plant genetic resources are a heritage of mankind and consequently should be available without restriction. The legally binding Convention on Biological Diversity, which came into force in 1993, encompasses not only plant genetic resources but all living organisms. In contrast to the FAO undertaking, the Convention affirms that States have sovereign rights over their own biological resources.

In January 2000, the legally binding Cartagena Protocol on Biosafety was adopted. This Protocol lays the foundation for a global system for assessing the impact of GMOs on biodiversity. It outlines the obligations of all countries that are signatories to protect biological diversity from the potential risks imposed by living modified organisms resulting from modern biotechnology.

At the heart of the Protocol is the concept of advanced informed agreement to the import of GMOs. This means that an exporting country must inform an importing country of its intention to export. It must also provide information on the GMOs, including an appropriate risk analysis. For its part, an importing country is required to give its informed agreement to the import. This implies that, although signatory states have the right to prohibit the import of GMOs if advanced informed agreement has not been given, they must also be capable of determining the validity of a safety assessment.

In addition, the Protocol advocates the "precautionary approach," described in the 1992 Rio Declaration on Environment and Development as a principle of international environmental law. The Protocol also proposes to establish a Biosafety Clearing-House for collecting, sharing, and disseminating information on risk assessment and management, which is particularly important for developing countries. Thus far, 81 countries have signed the Protocol, including Bangladesh, PRC, Indonesia, Republic of Korea, Malaysia, Philippines, and Sri Lanka.

Approaches in Selected OECD Countries

A sample of industrialized countries shows a diversity of approaches to food and environmental safety and in the agencies or ministries that play key roles in the national biosafety system. For example, the US has chosen to develop biosafety capacity within existing institutions (notably the Environmental Protection Agency and the Department of Agriculture). Denmark has designated the National Forest and Nature Agency as the lead agency for biosafety. In Norway, in accordance with the Gene Technology Act, the application of GMOs is subject to ethical and social conditions. A GMO can be approved for use only after the assessment of its benefit to the community and its contribution to sustainable development. In Australia, apart from the different ministries involved, an Advisory Committee on Genetic Engineering was created some years ago.

That will soon be replaced by an office with an enforcement rather than an advisory role. A public hearing must be held before authorization is given to release a GMO. Australia has recently created the Office of the Gene Technology Regulator to operate in the Commonwealth Department of Health and Aged Care.

Japan has based its risk assessment system on the principles and concepts developed by OECD. Different Japanese ministries and agencies have developed guidelines in accordance with their responsibilities on a product-sector basis.

A growing number of countries are introducing guidelines or mandatory measures for labeling genetically modified foods. In the US, formerly strongly opposed to labeling, the Food and Drug Administration has recently been charged with developing guidelines for voluntary labeling of food products to indicate whether or not they contain GMOs. In the European Union, food products containing more than 1 percent GMO must be labeled, stating clearly: "this product contains GMOs." The Australia New Zealand Food Standards Council is drawing up a new Food Standards Code. Labeling requirements will be drawn up in accordance with this joint Code. Japan is planning the introduction of mandatory food labeling in April 2001.

The OECD has also played an active role in promoting the harmonization of biosafety procedures among its member countries so that information and data gathered in the course of risk assessment may be mutually acceptable.

Emerging Regulatory Systems in Asia

A growing number have set up national biosafety systems; others are formulating national biosafety guidelines. Most countries have not yet completed guidelines for commercialization. The PRC is the only country in Asia where transgenic crops have been approved and released for sale. A number of countries are also in the process of introducing food labeling. Indonesia is planning to introduce mandatory labeling in April 2001.

ASEAN has undertaken a number of initiatives aimed at the harmonization of biosafety procedures among its member countries. In 2000 it also conducted a joint workshop with the Ministry of Science and Technology of the PRC to discuss issues related to transgenic plants. The Asia-Pacific Economic Cooperation has a number of capacitybuilding and technical cooperation activities in support of risk assessment, risk management, risk communication, and public acceptance. Responsibility rests with its Agricultural Technical Cooperation Experts Group and its Sub-Group on Research, Development and Extension of Agricultural Biotechnology.

Although some Asian countries have regulations in place, weak law enforcement sometimes results in the release of GMOs without proper procedures having been followed. For example, P.T. Monagro Kimia (a subsidiary of Monsanto) released Bt cotton in Indonesia without complying with Indonesian environmental law on safety procedures, environmental assessment, and information disclosure. Law enforcement in Asia needs strengthening to avoid risk to human and environmental health.

International and Donor Support for Biosafety Capacity Building in Asia

A growing number of initiatives are being taken to provide support for Asian countries in developing their capacities to safely manage biotechnology. A representative sample follows.

Following the signing of the Biosafety Protocol, UNEP, through its Global Environment Fund, is preparing a major capacity-building program for developing countries. It will also be responsible for the establishment of the Biosafety Clearing-House. FAO serves as the secretariat for the International Plant Protection Convention, which has recently established a working group on the Phytosanitary Aspects of GMOs, Biosafety and Invasive Species. UNIDO has conducted training courses in biosafety in a number of Asian countries. In collaboration with OECD, it has compiled an information source and database (BIOBIN), which provides information on biosafety methods and procedures, and on the status of the regulatory frameworks in individual countries.

OECD has established a network of international organizations with activities related to biosafety to enhance the exchange of information and to facilitate cooperation.

IBS provides a number of services to enhance developing country capacities to manage agricultural biotechnology. It gives short training courses annually for managers of agricultural biotechnology research programs in Asia. Biosafety is a key element in these courses. IBS also provides advice to individual countries.

USAID is involved in a number of biosafety training activities in Asian countries, some of which have been provided by its Agricultural Biotechnology Services Program. ISAAA is currently setting up a Global Knowledge Center on Crop Biotechnology based at ISAAA's SEAsia Center at Los Baños in the Philippines. ISAAA also organizes capacity-building workshops and training for regulatory officials and scientists from the Asian region.

Genetically Modified Organisms: Food and Environmental Issues

In industrialized countries, considerable progress has been made in methods, approaches, and experience in the safe management of GMOs. Overall these methods have proven to be effective, particularly with respect to food safety. The development of a transgenic soybean containing a protein derived from Brazil nuts, potentially useful in animal feed, was abandoned when the safety assessment revealed that the protein was probably an allergen. In response to consumer concerns, antibiotic markers are now being phased out as alternatives are being developed. That despite consensus among national and international regulatory authorities and the scientific community that the markers pose no threat to human health.

Despite acknowledged progress, areas of scientific uncertainty or disagreement persist. New challenges for risk/safety assessment, management, and monitoring will arise as a second generation of food, agricultural, and public health products emerges during this new decade. Not only are the ranges of organisms and numbers of traits expected to expand, so are the numbers and diversity of geographic and ecological sites in which they are released. These new products will not lend themselves easily to current approaches and methods to risk and safety assessment and to management. This highlights the need for constant review and improvement of assessment principles and procedures, and the urgent need for collection and analysis of ecological data. Other areas where consensus has not been reached among scientists themselves, or among scientists and policymakers, include (i) the precautionary approach as a method of dealing with scientific uncertainty, and (ii) methods for traceability.

The potential for the development of insect resistance to Bt transgenic crops is another area where the scientific community, industry, and environmental NGOs have not reached consensus. Risk management approaches at the farm level have so far been applied mainly on large-scale, commercial farms where monoculture is practiced. The same approaches may not be effective and, indeed, may not be necessary in the mixed farming systems of Asia.

Efforts to address the issue of resistance to transgenic crops have recently been initiated in Asia. A first International Consultative Workshop on Effective and Sustainable Use of Agricultural Biotechnology in Integrated Pest Management in

Developing Countries was held in November 2000. Organized by IRRI, Zhejiang University, and the International Organization for Biological Control, the workshop tried to determine strategies for the safe deployment of crops such as cotton and rice with novel Bt genes for pest resistance.

Among OECD countries, the continued monitoring of GMOs after release into the natural environment is also an area where opinions differ. On the one hand, it is argued that monitoring should be continued so that unforeseen risk can be managed and ecological impacts can be assessed.

On the other hand, it can be argued that the high costs of monitoring are not justified when transgenics have already successfully passed the hurdles of risk identification, safety assessment, and risk management.

It is clear that the potential risks of GMOs may vary by site and over time. Continued research will be necessary to improve data collection and analysis and to modify or devise new risk management strategies as needed.

Future Perspective

To date, a number of Asian countries have made substantial investments in biotechnology R&D initiatives. Some public research institutions have reached the stage of field testing of new crop varieties, major research projects are coming to fruition in IARCs in the region, and private companies are also conducting research in the region, or have products waiting to be commercialized and exported to the region. It will be necessary to have biosafety procedures in place to ensure that the benefits of these technologies are realized.

Although many opportunities exist, from donor and other development agencies, in support of capacity building for the safe management of biotechnology, and in setting up national systems, it is the responsibility of national governments to ensure that national regulatory systems are applied, enforced, and monitored. Setting up an effective national regulatory system for the safe management of biotechnology requires substantial resources, both financial and human. It is also a political process and, as such, requires political commitment.

The Cartagena Biosafety Protocol confers both rights and obligations on countries that are signatories. A country is free not to allow the import of GMOs on its territory. At the same time, it has an obligation to establish a national biosafety system.

This means that even if a country decides to ban the import or use of GMOs on its national territory, it will still need to comply with the obligation to develop a biosafety system for the control of such organisms.

It is anticipated that UNEP, the World Bank, and United Nations Development Programme (UNDP), through the Global Environment Fund, will contribute funding and technical assistance to developing countries for capacity building in biosafety. Even so, compliance with both the provisions of the Convention on Biological Diversity and the Cartagena Biosafety Protocol will place a heavy burden on some countries.

Apart from obligations under the Biosafety Protocol, if countries wish to control the import or introduction of GMOs on national territory, it will be necessary to have the means of detection in place. Similarly, from the perspective of international trade, when countries export agricultural or food products to countries that either prohibit the import of GMOs or require mandatory labeling for food or food products containing GMOs, the requisite regulatory procedures need to be in place.

The regulatory framework for biotechnology should not, however, be regarded in isolation from the broader policy context of agriculture and the contribution that biotechnology might make in the particular economic, social and environmental context of individual countries. Countries have a range of options in determining policies and priorities for biotechnology.

These might range between active promotion of biotechnology approaches, "wait and see," or rejection. Similarly, countries have a range of options in designing and implementing a national biosafety system geared to their particular technical, legal, and institutional realities. For example, a complex and time consuming regulatory system might impose prohibitive costs for local enterprises, particularly start-up companies.

PRC is the only Asian country so far with experience of managing and monitoring new transgenic crops in the field on a large scale. Some countries are gaining experience in field testing, while others have not yet completed biosafety guidelines. Information on socioeconomic and scientific developments in biotechnology and biosafety can be accessed from a number of information sources, including the Biosafety Clearing House mechanism to be provided under the terms of the Cartagena Biosafety Protocol. These sources are especially important for countries such as Cambodia, Lao PDR, and Viet Nam.

USING BIOTECHNOLOGY TO ADDRESS POVERTY REDUCTION AND FOOD SECURITY

Linkage Between Biotechnology and Poverty Reduction and Food Security

Agricultural biotechnology will contribute to poverty reduction and food security if scientists can develop technologies to increase quality and yields of food crops, and the technologies are adopted by small farmers. For this to happen, biotechnology R&D will have to meet four conditions:

(i) It must address both the problems of small farmers in rainfed areas where most of the poor live, and those of small farmers in irrigated areas, which provide the bulk of food grain supply in Asia.

(ii) It must focus on crops, livestock, and fish commonly grown by small farmers. Major crops are rice, tropical maize, wheat, sorghum, millet, banana, cassava, groundnut, oilseed, potato, and sweetpotato. Biotechnology R&D should also focus on high value cash crops (e.g., cotton, soybean, and vegetables) that can increase the incomes of small farmers through crop diversification. The prospect for improving these crops is bright due to the large demand for them in urban areas and in international markets. Fish and livestock (cattle, sheep, goats, pigs, and chickens) are also important.

(iii) The technology to be developed and delivered to small farmers must be simple, low cost, and carry little or no risks to human health and the environment. As in the case of the Green Revolution, the most effective strategy to increase food production is through improved seeds that possess high yield potential, fertilizer responsiveness, resistance to pests and diseases, good agronomic characteristics, and good nutritional quality.

(iv) Biotechnology development should be accompanied by favorable policy environment; good governance; investments in rural infrastructure; agricultural research and extension; and credit and marketing.

Why Invest in Agricultural Biotechnology?

In much of Asia. yields of major food grains are stagnant or declining in the face of population increases. Pests and diseases cause substantial preharvest and postharvest losses of crops, livestock, and fish. Solutions to many of these problems may lie in the various applications of modern biotechnology.

The use of molecular markers to tag specific traits is accelerating the breeding of new varieties of plants and animals. New understanding of plant and animal genes through genomics may offer ways of increasing crop yields.

These new developments when used in conjunction with developments in the physical and social sciences, offer more sustainable means for obtaining necessary productivity increases that are less dependent on environmentally damaging inputs of chemical fertilizers and pesticides. Given appropriate policies and necessary human and financial resources, modern biotechnology could make an extremely important contribution to future agricultural growth.

During the next 25 years, Asia will need a Second Green Revolution, often called Biorevolution or Doubly Green Revolution. Conway (1997) pointed out that the next technology-driven revolution must be doubly green—it must increase food production at a faster rate than in recent years without significantly damaging the environment. It must also increase incomes and increase access to food by the poor. The differences between the Green Revolution and Biorevolution are described in Table 5.1. Compared to the Green Revolution of the 1970s, Biorevolution will be characterized by the following features:

(i) Potentially many crops (particularly high value and specialty crops), will be affected as well as livestock and aquaculture.

(ii) Potentially all areas, both irrigated and rainfed, will benefit from biotechnology R&D.

(iii) Technology development and dissemination will substantially involve the private sector with the public sector playing the role of facilitator and regulator.

(iv) Many processes and products will be patentable and protectable.

(v) Capital costs of research will be high.

(vi) Molecular and cell biology expertise will be required in addition to expertise in conventional plant breeding and other agricultural sciences

Modern biotechnology (genetic engineering) is not a silver bullet for achieving food security, but used in conjunction with other techniques it may be a powerful tool in the fight against poverty and food insecurity (Persley and Lantin 2000). Other approaches are available and should be used. Narrowing the yield gap between those obtained from farmers' fields and those from experiment stations using the current technologies is just one example. However, there is concern that some conventional alternatives will not be able to produce the desired results within a limited time. The advantage of modern biotechnology rests on the speed at which desired crop varieties are produced. In some cases, the desirable genetic combination of traits is simply not possible through common breeding methods, and can be done only through genetic engineering. To increase food production by at least 40 percent within the next 25 years, Asian countries not only have to move toward the best technological frontier (to push farmers' yields to the optimum level), but keep moving the technological frontier itself. As long as product safety, environmental and ethical concerns, and IP issues are adequately addressed, modern agricultural biotechnology has the potential to significantly increase the quantity and quality of the food supply for developing countries.

Applying Biotechnology in Resource Poor Areas

The resource-poor, rainfed areas in Asia are home to many poor people, and their population is growing rapidly. Although migration may sometimes be the only viable livelihood strategy, in many such areas sustainable intensification of agriculture may be the best way to achieve poverty reduction and food security.

Development policy has often neglected these areas characterized by poor soils, shorter growing seasons, lower and uncertain rainfall, and little infrastructure or access to markets. Yield increases usually lag behind population growth. Efforts by poor farmers to expand cultivation onto new lands to eke out survival often lead to deforestation and erosion of fragile highland soils. That in turn threatens hydropower, road, and irrigation infrastructure in the lowlands. Policies and investments are required that can achieve food security in ways that protect natural resources, thereby breaking the vicious cycle of poverty, low productivity, and environmental degradation. Creation of nonfarm, rural jobs will also be important (ADB 2000b).

Sizeable investments in biotechnology in Asia have been made in rainfed areas, although DMCs continue to give high priority to irrigated agriculture. Irrigated areas produce 75 percent of total cereal production in Asia. Many IARCs such as the Asian Vegetable Research and Development Center, CIMMYT, CIP, ICRISAT, ILRI, and IRRI; and most DMCs have invested heavily in R&D on orphan crops such as rice, tropical maize, sorghum, potato, banana, groundnut, and sweetpotato.

Applicability of Biotechnology to Specific Agricultural Objectives

In Asia, biotechnology has been applied mainly to develop improved varieties adapted to specific environments. Specific areas where new applications of biotechnology could address poverty reduction and food security are summarized as follows:

(i) Increasing Productivity and Stability of Crops in Rainfed and Marginal Environments. Producing more food on the same area of cultivated land would reduce pressure to expand cultivated areas to forests and marginal areas.

Broadening tolerance of existing HYV cereals for drought, flooding, salinity, heavy metals, and other abiotic and biotic stresses would increase yields in rainfed areas. CIMMYT and IRRI, in cooperation with NARSs in different countries, are focusing their efforts to develop HYVs of rice and maize for rainfed areas.

(ii) Improving Water Use Efficiency in Crops. Future availability of water for agriculture is a major issue. By 2020, there may be a water crisis in Asian agriculture (ADB 2000c). Crops with higher water use efficiency and tolerance for drought would be an advantage. With the completion of the rice genome, scientists will be able to identify genes responsible for drought tolerance.

(iii) Integrated Pest Management (IPM) Strategies to Reduce Pesticide Use. The insecticides now used in many developing countries are often older, broad spectrum, acutely toxic compounds. Many are now banned in industrial countries except for export. They are a significant health hazard to farmers and farm workers throughout Asia. Over 80 percent of total pesticide use in Asia is on cotton, rice, and vegetables. In the past, governments encouraged the use of chemical pesticides as part of a yield-increasing package of farm inputs, and some governments still subsidize pesticides. The need is now recognized for more sustainable approaches to IPM to reduce losses to pests without harmful side effects on human health and the environment. New pest-and diseaseresistant crops, produced with the aid of modern biotechnology, may be important components of IPM strategies (Persley 1996).

Early applications are being seen in the use of novel sources of resistance (Bt genes) to control insect pests on maize and cotton. New pest-resistant cotton varieties are being grown widely in the PRC by 3 million farmers covering 500,000 ha (Pray 2000). Similar cotton varieties are undergoing field tests in India, Indonesia, and Thailand. Genetically modified maize resistant to insect pests is being tested in the Philippines.

(iv) Increasing Disease Resistance in Crops. Biotechnology applied to disease resistance in crops would be especially valuable for combating diseases in tropical environments that have not been controlled by conventional means. These include important diseases such as downy mildew of maize; bacterial blight and blast of rice, papaya ringspot virus; banana bunchy top virus; and bacterial wilt of tomato, eggplant, and potato.

(v) Increasing Nutritional Quality. Enhancing the protein, vitamin, and micronutrient contents of food grains would greatly benefit poor consumers who cannot afford supplementary vitamins and micronutrients. Under the Golden Rice Project, IRRI scientists are undertaking biotechnology research to enhance the vitamin A content of rice by introducing genes from daffodil (IRRI 2001).

(vi) Increasing Sustainable Production of Livestock. Research to increase the productivity and quality of farm animals, especially of dairy cows, small ruminants (sheep and goats), and water buffalo is promising. Vaccines and disease diagnostics for control of epidemic and endemic diseases of livestock are being developed.

(vii) Increasing Productivity in Fisheries and Aquaculture. Development of vaccines and disease diagnostics for control of epidemic and endemic diseases of shrimp in Thailand is a good example of where modern biotechnology has been used successfully.

The Success Story of Shrimp Biotechnology in Thailand

Shrimp is one of the top 10 exports for Thailand, generating about $1.5 billion in annual export earnings. In 1999, there were about 25,000 shrimp farms, producing 240,000 t valued at 87 billion baht ($1.8 billion), and employing about 130,000 people.

Diseases are a major constraint in the production of cultivated shrimp. In 1994, the white spot syndrome virus became a major disease in the PRC, and quickly spread to the rest of Asia, including Thailand. The disease caused a sharp drop in shrimp production in the PRC from 155,000 t to 35,000 t, a decline *of about 77 percent. National Center for Genetic Engineering and Biotechnology (BIOTEC) quickly supported research at the Universities of Chulalongkorn and Mahidol to develop DNA probe technology for the rapid detection of major shrimp pathogens. Rapid diagnostic reagents were quickly developed, the technology was transferred to the shrimp industry, and the virus was effectively controlled. It is estimated that shrimp biotechnology research and development has yielded a net benefit of about $1 billion since 1996. That constitutes a return on investment of 5,000 times in a single year, since the total cost of the biotechnology was only $200,000.*

Source: Morakot Tanticharoen (2000).

Benefits from Agricultural Biotechnology

Realized or potential benefits of agricultural biotechnology can be categorized as economic, social, and environmental. A large share of the benefits are concentrated in industrialized countries, where a diversity of applications are widespread in human and animal health care and in many aspects of food production and processing.

As the first transgenic biotechnology product was not commercialized until 1996, little quantitative analysis has been publicly available until recently. The studies published by ISAAA (James 1997, 1998, and 2000) have helped fill the gap, at least with respect to transgenic crops. ISAAA analyzed the global status of transgenic crops, including the area planted and the value of the market for transgenic seed, which was estimated between $2.7 billion and $3.0 billion in 1999.

The studies show that the economic benefits were fairly evenly distributed between the companies that developed the technology and the farmers growing transgenic crops. ISAAA studies have also demonstrated positive environmental impacts. In the cases of herbicide-tolerant soybean and herbicide-tolerant canola, herbicide use was considerably reduced even though weeds were better controlled, and yields increased significantly. In the cases of Bt maize, cotton, and potatoes, targeted insect pests were effectively controlled and yields were increased. Insecticide use for nontargeted insects was also reduced.

Economic Benefits in Developing Countries

Few published analytical studies have attempted to assess the impact of biotechnology in developing countries. Ex-ante studies (Qaim 1999) on transgenic pest-or virus-resistant sweetpotato in Kenya and on transgenic virus-resistant potato in Mexico suggest substantial benefits for both producers and consumers. They point out, however, that smallholders will fully realize benefits only with improved farm-level management and more efficient seed distribution.

The introduction of pathogen-free banana planting material in Kenya illustrates the benefits that accrue to small-scale farmers, mainly woman. Yield increases from the new material have been substantial; in some cases yields almost doubled. This example is significant in that it involved community groups and farmers in the development and field-testing. By the time the new planting material was available commercially, the benefits of the technology had already been demonstrated to farmers, with the result that initially demand exceeded supply. The project won the science and technology prize in a Japan/World Bank-sponsored network of environmentally beneficial development projects. The Kenyan experience is highly relevant to Asia.

Benefits in Asia

A number of biotechnology applications are widespread in Asia. These include biofertilizers and biopesticides, and products from tissue culture and micropropagation (banana, coconut, maize, potato, cassava). They also include diagnostics, vaccines, and embryo transfer for livestock. An ex-post evaluation on the impact of a transgenic crop at the farm level in Asia analyzed the impact of Bt cotton in the PRC (Pray et al. 2000) A sample of 283 cotton farmers from five counties of Hebei and Shandong provinces was studied. The key findings were (i) the cost to produce 1 kg of cotton was reduced by 20-30 percent, depending on the variety and site, and (ii) net income and returns to labor of all the Bt varieties were superior to the non-Bt varieties. Smaller farms and those farms that had lower incomes consistently derived larger increases in net income than larger farms and those with higher incomes.

The use of Bt cotton substantially reduced farmers' use of pesticides, usually from 12 sprays per season to 3 or 4. The study also found some preliminary evidence of positive impact on farmers' health. An ex-ante study on the impact of the introduction of biotechnology for the reduction of corn-borer infestation in the Philippines (Gonzales 1999) has postulated three kinds of benefits. First, the introduction of transgenic Bt maize would enhance the competitiveness of the Philippines from the point of view of import substitution and export prospects. Second, it would increase farmers' incomes. Third, it would reduce pesticide use.

Weighing Risks and Benefits of Biotechnology

The evidence thus far available suggests potential economic, social, and environmental benefits from the use of biotechnology. It also Biotechnology to Address Poverty Reduction/Food Security Problems 61 suggests benefits for both food and export crops, for commercial and small farmers, and for consumers in the form of nutrition-enhanced products. The potential benefits will be realized only if a number of conditions are met:

(i) Public institutions will need to play a key role in directing the research agenda toward social and equity goals.

(ii) Complementarity between the public and private sectors, in both R&D and in technology development and delivery, needs to be maximized so that public investment and research is focused on those areas not of interest to the private sector.

(iii) Governments must be committed to developing technologies for the public good to ensure that the poor, including women, have access to the new technology.

The benefits and risks of biotechnology need to be assessed on a case-by-case basis. The uncertainties and the risks are yet to be fully understood, and the benefits are not yet fully exploited. It seems important not to deny people access to new biotechnology, so long as they are fully informed of the potential risks and benefits in making their choices. We have an ethical imperative not only to keep the technology portfolio open to biotechnology and genetic engineering, but also not to lose time. Every minute lost, every decision delayed, means more deaths from starvation and malnutrition (Leisinger 2000).

ISSUES AND OPTIONS

Policy and Priority Setting

Issues

A surprising amount of biotechnology R&D is already being done in Asian developing countries. There have been few studies, however, on the cost-effectiveness of biotechnology R&D in these national programs, and whether priorities are research-driven or demand-driven. On the other hand, countries have a justifiable concern that they will be left behind if not involved in the technology. In establishing policies and priorities, Asian countries should consider that (i) biotechnology requires a high level of capital and specialized technical skills, (ii) the public sector is constrained by inadequate capacity, (iii) the private sector has little interest in investing in technology for poor farmers, (iv) there are potential risks associated with GMO development, and (v) it is difficult to establish an effective biosafety framework. Some of the broad policy issues relate to capture of benefits. How can the benefits of biotechnology reach poor farmers? That is, how can biotechnology programs be better linked to extension programs? Who will be winners and losers from genetically engineered crops? Will there be a widening income gap between poor farmers and rich farmers who have access to credit to purchase biotechnology-derived seeds?

Options

There is a need to assist DMCs in policy and priority setting. Policy setting must be based on informed decisions on appropriate choices regarding the use of new biotechnology applications in agriculture and their prospects for enabling sustainable productivity. The likely impact of biotechnology on poor people, either directly by increasing crop yields and farm incomes or indirectly by improving their environment, should be considered. There is a need to encourage regional and national coordination of policy on biotechnology (especially of GMOs), and consistent distribution of responsibilities between agencies in different countries. That may be difficult to achieve, but the problems of poor coordination have been highlighted by the European experience (Levidow et al. 1999). If Asian countries are to participate more fully in biotechnology, they will also need to expand their own capacities to undertake biotechnology research linked to the problems of smallholders and marginal farmers.

Priorities for Research and Development

Issues

Countries have a range of options in determining policies and priorities for investment in biotechnology research. These might range from active promotion of biotechnology approaches, to a wait and see, stance, or to rejection of biotechnology R&D altogether. Whatever the case, technical and policy capacity will still be needed to support international trade and biosafety agreements. Donors, too, have a range of options. Should they focus their biotechnology investment on the more advanced developing countries, where there is more capacity and more likelihood of using the results? In any event, countries need to establish planning for investments based on demand. For example, is excessive emphasis given to plant biotechnology at the expense of animal biotechnology— especially given that the livestock sector is much faster growing?

There is a need to establish where biotechnology methods could deliver varieties that conventional breeding cannot. Where either a conventional or biotechnology-based breeding approach could be used to target a particular characteristic, there is the need for comparative economic and risk analyses of the alternative approaches. That should also take into account that many genetic engineering approaches also require conventional crossing and selection to transfer the desired character from the transformed plant into an adapted, elite variety. Many early biotechnology research programs in Australia made the mistake of operating in isolation from breeding and agronomic improvement programs, hindering application of the research. It is important that developing countries do not repeat the mistake.

Options

Obtaining a clear view of which crops and characters should be targets for genetic manipulation in Asian developing countries is a critical step, but it has rarely been addressed systematically

(Woodend 1994). Research should focus on crops relevant to small farmers and poor consumers in developing countries: cassava, yam, sweetpotato, rice, maize, wheat, millet, and possibly papaya. The limitation is that relatively little biotechnology research is being undertaken on many of Asia's basic food crops or on the problems of small farmers in rainfed and marginal lands. In many cases, Asian countries will need to expand their own national and regional capacity to undertake biotechnology research to address these problems. Once a clear view is obtained on which crops and characters to target, it may make more scientific and economic sense to purchase, license, or import particular elements of technology.

Indeed, there have been a number of successes of crop genetic engineering in developing countries with technology already commercialized or in field trials. These include: insect resistance in maize and cotton and several horticultural crops using genes from Bt and other sources; resistance to a range of viruses in potato (viruses x, y, and leafroll); and ringspot virus resistance in papaya and cucurbits. Rice with resistance to bacterial blight and with increased content of beta-carotene or iron has been produced, as well as tomatoes with elevated lycopene (related to vitamin A).

Support of agricultural biotechnology research is important to developing countries for several reasons. Apart from direct benefits from applications, developing country ownership of IP can be traded for other technology. The existence of active programs in-country can provide a balanced understanding of issues by technocrats who make policy in areas such as biosafety. Some other R&D needs include the following:

(i) One of the most profitable applications of crop biotechnology in the short term to medium term will be through better implementation of molecular markers for selection in conventional breeding programs. That will require closer linkages between biotechnologists and plant breeders in many institutes.

(ii) Engineering of desirable traits in several of the major cereal crops of Asia (e.g., rice, maize, wheat) is being addressed by some of the international networks and through programs at IARCs. Developing country NARSs need to strengthen their ability to adopt the technologies and the germplasm as they are developed by these international initiatives.

(iii) It will be important to identify specific orphan crops and animals with short-to medium-term potential for application of existing biotechnology in developing countries. In some cases, this may best be done through developing more efficient methods of crop improvement or through the use of MAS for complex traits. In others, the most appropriate approach may be through the development of transgenic varieties with specific characteristics such as disease resistance or improved nutritional quality.

(iv) There is a need to ensure that the genomic knowledge of the major food crops in Asia is obtained and is available in the public domain as the basis for the development of new crop varieties for the public good. Such knowledge will underpin the development of the new food crop varieties for Asian agriculture over the next decades. The rice genomics program would be a good model for others (Sasaki 1999).

(v) Apomixis in a wider range of crops has huge potential for simplifying the breeding of adapted genotypes, preservation of hybrid vigor, and improved propagation of crops that currently rely on vegetative propagation. It is also an important tool for the discovery of genes that are potentially important in stress tolerance.

(vi) There is a need to build the capacity of Asian developing countries to undertake laboratory and on-site testing of GMOs. An increasing number of published protocols are available as well as simple, commercial test kits.

Intellectual Property Management

Issues

Because the private sector invests heavily in, and holds many of the advanced biotechnologies, new discoveries in biotechnology may be protected by plant variety protection, patents, or trade secrets. This raises the issue of IPR. The 1995 Agreement on Trade-

Related Aspects of Intellectual Property (TRIP) requires all countries to provide some sort of protection for plant varieties. TRIP requires all signatories to extend IP protection to microorganisms, plant genetic material, and techniques used for genetic manipulation. Although plants and animals, other than microorganisms, may be excluded from patent protection, countries are required to provide protection either by patents or by an effective sui generis system (see details in Appendix 13). Strengthened IPRs will increase the flow of technologies and products from developed to developing countries, and provide new incentives for local research and innovation. But one fear is that industrial nations, using genetic resources originating from developing nations, could develop GMOs or techniques and then restrict developing nations' access to the technology by employing the IPR provision of TRIP. Asian countries need to (i) develop a policy toward IP to protect their own discoveries, (ii) develop an effective approach to cooperation with the private sector in R&D, and (iii) encourage private sector research on products to benefit small farmers in Asia.

IP protection for plant genetic resources are being considered in the context of the Convention on Biological Diversity and renegotiation of the International Undertaking on Plant Genetic Resources of FAO. The issue has also been discussed in the World Intellectual Property Office in the context of traditional resource rights. In addition, environmental aspects of access to genetic resources are being discussed in UNEP and World Trade Organization (WTO). All these discussions are intended to ensure that existing international agreements encourage biotechnology development relevant to Asia's poor while protecting IPR of indigenous peoples. Most Asian countries are members of the World Intellectual Property Office, have a national patent regime, and are signatories to the Patent Cooperation Treaty and to the Paris Convention. With the exception of the PRC, whose application is pending, most are also members of WTO.

In conformity with TRIP, a number of countries are developing plant variety protection laws and are in the process of joining the International Convention for the Protection of Plant Varieties. These include India, Indonesia, Malaysia, Philippines and Viet Nam. Some countries (Bangladesh, India, Philippines) are currently formulating laws for the protection of biodiversity and community knowledge.

Proprietary interests in biotechnology affect developing countries in several ways, requiring them to consider issues relating to the (i) *licensing in* of technology owned by developed countries or multinational companies, (ii) their *freedom to operate* with their own inventions, and (iii) their ability to commercialize or export products (Maredia et al. 1999). Changes in the last couple of decades in the IP environment in which agriculture operates include the patenting of living organisms in some countries, and establishment of plant variety rights. The protection of target genes and of enabling technologies (e.g., gene regulation, markers for selection, promoter sequences, and transformation technologies) mean that developing country organizations have to be conversant with licensing activities.

Among them are technology use agreements, material transfer agreements, and commercial licenses (Mascarenhas 1998). Some alternative strategies to gain freedom to operate can include: (i) inventing around current patents, (ii) redesigning constructs to synthesize genes to reduce reliance on external technical property, (iii) asking IP owners to relinquish claims or provide royalty-free licenses, (iv) ignoring all IP and technical property; or (v) seeking licenses for all IP and commercial property, which is certainly the safest route to building public-private sector cooperation. Many developing countries, however, use such *inputs* without formal permission.

This issue becomes important at the time of commercialization of a technique or process, or release of a variety. By then, it is often too late to change direction or incorporate an alternative piece of technology in a research or breeding program. A recent ISNAR survey of five Latin American countries showed that NARS laboratories used proprietary technologies widely, but in many cases without formal agreements. Most of the proprietary technologies related to plant genetic engineering. They included markers for selection, transformation systems, promoters, and specific genes as well as marker techniques and diagnostics. There was also a lack of knowledge regarding IPRs in both academic and administrative ranks (Salazar et al. 2000).

IP management of biotechnology inventions can be very complex. Kryder et al. (2000) recently reviewed the example of Golden Rice, which is high in pro-vitamin A (beta-carotene). This rice is potentially valuable for those who are too poor to obtain green vegetables. A very large number of proprietary technologies were used to develop Golden Rice. It is a multitransformant, since three genes were introduced into a carotenoid biosynthetic pathway to produce the high beta-carotene levels. Other proprietary technologies included three transformation vectors: *Agrobacterium* transformation, plant regeneration, and DNA amplification. Thus, a country or organization that requires this technology must obtain consent from a wide range of partners.

Options

The IPR issue is complex and yet many Asian institutes have focused training only on the technical aspects of biotechnology research, rather than on other areas such as licensing and IP management. Steps required to obtain consent from a wide range of partners for a given technology need to be defined on a case-by-case basis. Therefore, training in IP regulation to facilitate technology transfer in evolving IPR systems, and assistance in developing a clear understanding of TRIP agreements is important. Systems are needed to enable countries to protect their own technology, while minimizing barriers that could hinder technology transfer between countries The adoption of IPR law by developing countries may make proprietary biotechnologies (seeds and planting materials) more costly. To overcome this problem, the public sector would need to consider buying exclusive rights to newly developed technology and make it available free to small farmers. Alternatively, the public sector and IARCs could invest in biotechnology R&D so that the resulting technology could be made available free to farmers.

Public-Private Sector Partnership

Issues

Significant increase in biotechnology investment by governments and donor agencies is crucial for achieving food security and poverty reduction. As mentioned earlier, there is insufficient profit motive to induce the private sector to undertake the R&D needed by small farmers. As Byerlee and Fischer (2000) point out, there have been significant market failures in applying biotechnology for the benefit of developing country farmers; therefore the public sector will continue to have an essential role. Because many of the techniques and products of modern biotechnology are privately owned, public agencies need modes of action to operate in an increasingly private sector world. Furthermore, both the public and private sectors have complementary assets needed for biotechnology to be applied to its full potential.

Many of the results of biotechnology research will be most easily transferred to poor farmers as seeds in much the same way as the results of the Green Revolution. But this time it is expected that the private sector will have a significant role in dissemination. The private sector, especially seed companies, is becoming more important in several Asian developing countries such as India and the PRC. Similarly, in Europe and North America several seed companies have formed alliances with global life science companies. Several Asian agrochemical and floriculture companies are involved in joint ventures with developed country companies. Thus, in line with ADB's Private Sector Development Strategy (ADB 2000), there is a need to increase the partnership between the public and private sectors.

Options

Present involvement of companies in agricultural biotechnology in developing countries in Asia (especially in major crops and large countries) suggests that the motivation is purely commercial. For example, seed for up to half the engineered cotton grown in the PRC has been commercially obtained from Monsanto. In many cases new mechanisms will be needed by which the private sector can assist with technology transfer, extension, and distribution of biotechnology products.

That may require some innovative arrangements involving special funds and other financial incentives for private companies. Multinationals also have an incentive to involve themselves in developing country partnerships/philanthropy to counter negative public impressions. Segmentation of markets will be important to allow developing country farmers to access the products of biotechnology under realistic conditions.

Brokering groups such as ISAAA, a USAID program, and CGIAR centers have already implemented a range of approaches. These include market segmentation based on crop and growth region, country income level, trade status, or crop variety. In other cases the public sector could offer to buy exclusive rights to newly developed technology and make it available either free or for a nominal charge to small farmers (but on a profit basis to commercial farmers). The private research agency would bear the risks, as it does when developing technology for the market (see details in Appendix 13). The other possibility is for the public sector agency to finance private R&D on orphan crops through competitive bidding.

There is a need to establish whether international organizations and funding agencies should play a brokering role in the dissemination of agricultural biotechnology. A neutral broker can have advantages. For example, ISAAA claims that its comparative advantages are cosponsorship by public and private sector institutions, independence, and neutrality (lack of financial interest in the technology). Funding agencies (e.g., USAID) have also served as brokers between public agencies and private companies. Centralization of technology transfer offices for agricultural biotechnology in individual countries would facilitate brokering agreements, and inspire greater confidence in the private sector.

Other biotechnologies such as micropropagation of plants through tissue culture, and production of biofertilizers, biopesticides, and selected vaccines may be suitable for small-and medium-sized private companies in developing countries. Such companies could take technology directly from local research institutes. There are many successful examples of this already. Such enterprises can create employment in rural areas and deliver affordable and useful products to farmers.

Biosafety

Issues

Biosafety concerns relate either to food safety and human health or to the environmental impact of genetically modified crops. There is also a broader ethical concern as to whether genetically modified foods are unnatural, and whether the use of the technology encourages the narrow control of agriculture by a few (multinational) corporations. One of the key principles in biosafety guidelines is that of *substantial equivalence* as the approach to identify differences between biotechnology-derived and traditional foods (Miller 1999). The main environmental risks relate to the (i) potential loss of genetic diversity in cropping systems; (ii) potential transfer of genes from herbicide-resistant crops to wild relatives, creating superweeds; (iii) ability of herbicide-resistant crops to act as weeds in rotation crops; (iv) escape of transgenes, especially antibiotic resistance markers to soil bacteria; (v) vector recombination to create new viruses; and (vi) with Bt toxins, insect resistance to the toxin and the effect of the toxin on nontarget organisms.

The need for developing countries to have functioning biosafety systems has strengthened since the adoption of the Cartagena Protocol on Biosafety in January 2000. The Protocol establishes a framework for regulating international trade in transgenic crops. Three major components of the Protocol have implications for individual countries as users, developers, and exporters of GMOs:

(i) International shipments that "may contain" transgenic food products must be so labeled. This applies only to large-scale shipments; it does not affect labeling requirements on consumer products, which are determined by each country.

(ii) Governments may use the *precautionary principle* to bar import of a transgenic product even in the absence of conclusive evidence that the product is not safe. The Protocol does not, however, override other international agreements, including WTO, which requires that import decisions be sciencebased.

(iii) To assist countries in making import decisions, a database will be established to make available uniform information on transgenic crop varieties.

The need for harmonization of regulations is not just an issue for developing economies in Asia, but also for developed countries. For example, Levidow et al. (1999) report on how difficult it has been for European countries to harmonize regulatory criteria despite an expressed position of using science-based criteria. Countries disagree on the amount of scientific evidence required to resolve uncertainties such as in the recent approval processes for Bt insect-resistant maize and herbicidetolerant canola.

Options

Countries have a range of options in designing and implementing national biosafety systems geared to their technical, legal, and institutional realities and in adapting the system to local needs, priorities, and capacities. There is a need to assess the adequacy and effectiveness of biosafety procedures for the testing, release, import, production, and use of GMOs in Asian countries.

Regulatory systems must be flexible to allow for either increased scrutiny or relaxation of controls (e.g., for containment of field trials) based on available scientific evidence. Harmonization of regulations between agencies is important. But it may be simpler and less expensive to embed biosafety regulation within existing institutions rather than to build new ones. Australia's approach has been to develop national policies and regulations for trade in GMOs within the country's existing regulatory framework. Thus, governments could be helped to build the infrastructure to manage GMOs within the context of developing a quarantine policy, managing sanitary and phytosanitary issues, and assessing risk and environmental impact. Collaboration between quarantine/ regulatory officials and environmental policymakers should be strengthened.

An area of potential concern is the lack of verified ecological data on the effects of genetically modified crops under Asian or tropical conditions. Governments and international agencies should be supported in participatory field studies on the ecological impact of first-generation of genetically modified crops such as insect-resistant cotton and rice. These assessments should involve local communities in the evaluation of the new technologies.

The likelihood that Asian developing countries and their agricultural export customers will adopt the labeling of food containing genetically modified products needs to be assessed. Several countries have introduced labeling systems or are considering them. In some cases, industry has expressed concerns about the extra costs involved in testing, ingredient tracing, and labeling. Several options for labeling have been proposed. They include (i) labeling based on the presence of detectable transgenic DNA or protein only; (ii) special labeling for GMO-free foods; (iii) labeling all foods derived from GMOs; and (iv) labeling foods or any food ingredients produced with GMOs (e.g., meat from animals fed with transgenic crop residues). It will be important to establish the impact of such labeling on the production, distribution, marketing, and exports of genetically modified foodstuffs.

Fostering Better Public Awareness

Issues

The future of biotechnology lies in public awareness and acceptance; good technology alone is not enough. For example, although nuclear power provided low-cost electricity for many countries, there has been substantial disadoption in recent years. The news media in several Asian developing countries regularly report on biotechnology and the controversy surrounding GMOs. Some aspects of the debate are similar to 74 Agricultural Biotechnology, Poverty Reduction and Food Security that in Australia, in which the controversy is seen as one of technocrats encouraging the use of biotechnology versus the average person who is unsure of the benefits. European countries, especially, are seen to have the luxury of excess food production. The need to alleviate poverty or enhance food security is not part of their debate. Certain additional issues appear in some developing countries. Some countries such as Thailand are major exporters of processed foods. They are concerned about the attitudes of their export customers toward genetically modified foods.

Options

Inclusion of public awareness activities from the onset of a biotechnology program, rather than at its completion, can greatly assist in gaining acceptance. For example, before starting development of a transgenic, insect-resistant, tropical maize in Kenya, CIMMYT consulted with NGOs, farmer organizations, media, and other stakeholders to create a supportive environment. Public awareness programs should be included in the activities of major institutions carrying out biotechnology R&D. That is already done in Thailand and the Philippines, using simple messages in local languages. Again, it will be equally important to involve community groups in the debate, not just scientists or the food industry. Finally, multinational organizations such as Monsanto have recently realized the necessity for developing a strategy to ensure informed public debate about the risks of biotechnology to human health and the environment, and to ensure consumers are fully aware of the true benefits and costs of biotechnology. Such campaigns must recognize that freedom of choice is valid.

CONCLUSIONS AND RECOMMENDATIONS

During the next 25 years, Asia will face a serious challenge on how to reduce poverty and achieve food security due to (i) an absolute increase in population; (ii) the doubling of its urban population; (iii) continued deterioration of water, forest, and soil resources; and (iv) the need to produce the food where it is consumed, because the share of total grain production traded has remained stable at about 10 percent. The most attractive strategy to meet this challenge is to increase smallholder agricultural productivity. This strategy will not only increase food supplies, but it will also increase smallholders' incomes, reduce malnutrition, and improve the livelihoods of the poor. Increasing smallholder productivity on a sustainable basis is complex. The challenge has to be addressed by modern science since the Green Revolution has already run its course in much of Asia. And it has also bypassed the rainfed and marginal area where most of the poor are concentrated.

Modern biotechnology brings new possibilities for achieving the sustainable increases in agricultural productivity that will be necessary to meet the projected demands for food by Asia's growing population. It has the potential to increase agricultural production and improve processing. Past large-scale investments in biotechnology have resulted in modest increases in crop yields, reduced dependence on pesticides, and better quality food. That could be followed by larger, dramatic increases in crop and livestock productivity; control of major diseases in livestock and fisheries;

increased resistance of crops to drought, salinity, and acidity; and new types of processed foods. Using modern biotechnology, new HYVs can now be developed much more quickly with greater precision compared with conventional breeding methods.

During the past decade there has been rapid progress in the application of modern biotechnology in developed countries, thanks to massive investment of the private sector. During 1996-2000, the total area planted to genetically modified crops expanded rapidly from 1.7 million ha to 44.2 million ha. Seventy-five percent were in the United States, with the remainder in other countries in Europe, Latin America, and Asia (mainly the PRC).

Some Asian countries, notably PRC, India, Indonesia, Malaysia, Pakistan, Philippines, Thailand, and Viet Nam, have already made modest investments in agricultural biotechnology. Their capacity to carry out biotechnology R&D has been strengthened with financial support from ADB, AusAID, USAID, the Rockefeller Foundation, the World Bank, and other funding agencies. The PRC has developed and released to farmers a number of transgenic crops, which now cover at least 500,000 ha. Some IARCs have also made modest investments in biotechnology to develop HYVs of orphan crops (e.g., tropical rice, tropical maize, sorghum, groundnut, and chickpea) in collaboration with NARSs. The development of safe and efficient biotechnology in Asia has been constrained by a shortage of trained personnel, lack of capital, poor management of IPR, ineffective biosafety regulations and enforcement, and the widespread perception of some NGOs that biotechnology poses serious risks to human health and the environment.

The risks of biotechnology to human health and the environment are confined to transgenic crops and livestock or GMOs. Other components of biotechnology such as microbial fermentation, tissue culture, marker-selected breeding, and disease control are relatively safe and have no adverse impact on human health and the environment. Biotechnology consists of a gradient of technologies ranging from simple, low risk technologies to complex, expensive, and highly risky technologies. Asian governments therefore have a choice of technologies to invest in depending on the availability of human and financial resources, and their capacity to monitor and evaluate potential risks. Agricultural biotechnology is not the sole means for achieving food security. But in conjunction with complementary activities, it may be a powerful tool in the fight against poverty. These complementary activities would include a favorable policy environment; good governance; investments in rural infrastructure, agricultural research, extension, and agricultural credit; and marketing.

Recommendations

General Strategy

To ensure that agricultural biotechnology will contribute to reducing poverty and improving food security in Asia, biotechnology R&D should do the following:

(i) Address the problems of small farmers in the rainfed and marginal areas where most of the poor live, yet not neglect the problems of small farmers in the irrigated areas.

(ii) Focus on economically important orphan crops, high value crops, and livestock to increase their productivity.

(iii) Develop low cost, appropriate technologies for small farmers, particularly the development of HYVs adapted to the rainfed and marginal areas.

(iv) Develop, test, and release technologies that will pose minimal or no risks to human health and the environment.

(v) Strengthen the extension, delivery, and regulatory systems to ensure that improved varieties and technologies will be disseminated widely to small farmers with little or no risk to consumers or the farmers themselves.

Role of Government

To use agricultural biotechnology safely and effectively for the benefit of small farmers in Asia, governments in the region should:

(i) Demonstrate a strong commitment to agriculture and rural development by providing adequate budget and staffing to the sector in general and agricultural biotechnology in particular.

(ii) Establish clear polices and priorities in biotechnology R&D to ensure that it can contribute effectively and safely toward poverty reduction and food security.

(iii) Enhance cooperation with the private sector in the development of biotechnology that will benefit small farmers.

(iv) Set up effective biosafety regulatory and enforcement systems to ensure that the risks of biotechnology (particularly those of genetically modified crops and livestock) will be minimized.

(v) Enact IPR laws that will protect and stimulate private sector investments in biotechnology in the region.

(vi) Organize dialogue with NGOs, consumers, and farmers on the benefits, risks, and opportunities in the use of new biotechnology.

(vii) Seek assistance from international organizations and funding agencies on specific problems in biotechnology that cannot be addressed using their own resources.

Suggested Policy for ADB on Agricultural Biotechnology

The major conclusion of this study is that funding agencies, including ADB, would be wise to continue and increase their investments in the safe applications of biotechnology, as one means to achieving poverty reduction and food security in Asia over the next 25 years. Achieving these goals with presently available technologies will be difficult given present trends and future challenges facing the rural sector in Asian environments. ADB can support biotechnology R&D in Asia through loans and technical assistance to both public and private sector entities. Accordingly, it is recommended that ADB consider the following measures: Recommendation 1. ADB should assist DMCs in policy and priority setting to enhance investments in the safe application of biotechnology.

(i) Provide information to enable governments to make informed decisions in relation to the use of new biotechnology applications in agriculture. Particular care needs to be taken to select those interventions that impact poor people, either directly by increasing incomes or crop yields and quality, or indirectly by improving their environment and prospects for sustainable productivity.

(ii) Assist DMCs in identifying R&D areas where the use of new scientific developments may help achieve breakthroughs in dealing with previously intractable problems. These may include ways to increase sustainable productivity in rainfed areas, for example by the use of molecular methods to select new crop varieties with higher water use efficiency.

Recommendation 2. ADB should increase dialogue with its DMCs in identifying potential benefits and opportunities in the use of different biotechnologies to address specific targets.

(i) Increase policy dialogue with governments on the importance of the rural sector in underpinning social, economic, and environmentally sustainable development. The important role scientific developments can play in making the rural sector more productive and more sustainable through better management of natural resources is an important issue. Good governance and beneficiary participation will ensure the poor, including women, have access to the new technology.

(ii) Support risk/benefit analyses as a basis for choices on the merits of new technologies to address particular problems relative to existing technologies and other options. An example of where risk/benefit analyses could be undertaken in the short term is on the potential use of transgenic cotton varieties in Asian cotton-growing countries, based on the experience of the use of transgenic cotton varieties in the PRC. The analyses could compare various technology options, including the use of pesticides and integrated pest management systems.

Recommendation 3. ADB should strengthen risk assessment and management capabilities in its DMCs through systematic capacity building.

(i) Assist smaller Asian countries to set up national regulatory systems appropriate to their size and resources, while being consistent with international best practice.

(ii) Support regional harmonization efforts and activities being undertaken through ASEAN and APEC for the development of agreed upon biotechnology standards, guidelines, and regulations.

(iii) Support the development and implementation of protocols to monitor the long-term ecological impact of GMOs in the environment. Initial priorities could be for monitoring the performance of transgenic cotton and rice, since cotton is already under widespread commercial cultivation in the PRC and rice is the major food crop in Asia.

(iv) Facilitate the evaluation of potential new products in the regulatory pipeline in several countries through small-scale trials conducted under international best practice guidelines. These experiments would give data that could guide risk/ benefit analyses and future decisions on the safety of specific applications of new biotechnologies in Asian environments.

(v) Strengthen law enforcement in most DMCs to ensure that the introduction and release of GMOs follow government environmental requirements and procedures.

(vi) Support ex-ante and ex-post studies on the socioeconomic impact of transgenic crops in the region and their impact on the poor, including women.

Recommendation 4. ADB should facilitate access to proprietary technologies and encourage greater private and public sector cooperation in the development and delivery of new products at affordable prices for the poor.

(i) Facilitate negotiations with private companies on behalf of developing countries and the international agricultural research community to access key enabling technologies potentially useful on orphan commodities in Asia (i.e. staple food crops, livestock, and fish consumed locally).

(ii) Act as an honest broker or as a convenor in facilitating more public-private sector partnership in research, development, and dissemination of technologies relevant to the needs of small farmers.

(iii) Examine the feasibility of providing incentives for local private sector development in areas that would develop biotechnology-based businesses in rural areas, with high potential for stimulating income and employment opportunities.

(iv) Examine the feasibility of providing incentives for multinational companies to develop products to benefit poor people. Such incentives may be provided through tax breaks, contract R&D, guaranteeing a base market for a successful product, strategic alliances, or underwriting joint ventures with national companies. There are some experiences with these types of incentives in the health sector that merit closer examination for their applicability to the rural sector.

(v) Support national capacity building in IP management and technology transfer.

Recommendation 5. ADB should support a strategic R&D agenda and associated human resources development in Asia to generate new knowledge and disseminate the results for the public good. It should support and fund national governments and IARCs to undertake important initiatives that will have significant impact on poverty reduction and food security in the long term in areas of market failure where the private sector is unlikely to invest.

(i) Applications of Biotechnology to Orphan Commodities. Support public R&D on orphan commodities by national and international agencies. This may best be done by developing more efficient methods of crop improvement through the use of marker-assisted selection for complex traits such as drought tolerance. In other cases, the most appropriate approach may be the development of transgenic varieties with specific characteristics such as disease resistance or improved nutritional quality.

(ii) Ecological Research. Support national governments and international agencies in developing methods for undertaking participatory field studies on the ecological impact of the first generation of genetically modified crops. For example, insect-resistant cotton and rice are being released for field testing and commercial production in

Asia, particularly in the PRC. These assessments should involve local communities in the evaluation of the new technologies, similar to the approaches developed in Asia for integrated pest management.

(iii) Strategic Research on Functional Genomics. Support regional efforts on functional genomics to understand the genetic basis of the agriculturally important crops and livestock in Asia. For example, genomic studies on rice, maize, and sorghum may identify potentially useful genes for drought and salinity tolerance that may have wide applicability across all cereals.

(iv) Exchange of Information. Support the sharing of knowledge and experience among Asian countries on the applications of biotechnology to specific targets, the risk/benefit analyses that underlie particular choices, and the data on food safety and environmental risks that will be acquired as experience accumulates and strategic R&D is conducted.

10

Agricultural Biotechnology for Africa

Few would disagree that the many claims and counterclaims concerning what biotechnology can or cannot do to solve Africa's food insecurity problem have mainly been made by non-Africans. It is no wonder that Florence Wambugu's (1999) excellent article titled "Why Africa needs agricultural biotech" is now widely cited by those who support the view that developing countries, particularly in SubSaharan Africa (SSA), stand to gain the most from modern biotechnology applications. The article explained in a nutshell some of the potential benefits Africa stands to gain by embracing biotechnology. Although opinions differ regarding the role biotechnology can play in African development, all (hopefully!) must agree about the urgency to eradicate the perpetual cycle of hunger, malnutrition, and death in a world of plenty.

It is an acknowledged fact that Africa is endowed with tremendous natural (including genetic) and human wealth that has yet to be harnessed to the benefit of its people. Sadly, some of this reservoir of resources have been disintegrating and the trend is bound to accelerate unless urgent measures are taken to stop and reverse this drift. Since farming is the most important source of income and sustenance for about three quarters of the population of SSA, there is no doubt that agricultural biotechnology (agbiotech) can make very substantial contributions toward increasing food production by rural resource-poor farmers, while preserving declining resources such as forests, soil, water, and arable land (Bunders and Broerse, 1991). However, application of modern biotechnology tools is not likely to significantly reduce the contributions that conventional disciplines such as soil science, breeding, plant health management, agronomy, agricultural economics, and social sciences make to enhance crop production.

In villages, constraints to crop production include pests, diseases, weeds, environmental degradation, soil nutrient depletion, low fertilizer inputs, inadequate food processing amenities, poor roads to markets, and general lack of information to make science-based decisions that underlie farming methodologies and systems. For some of these constraints, biotechnology is the most promising recourse to alleviate them. For example, an insect known as *Maruca* podborer is the major constraint restricting increased grain legume production in Africa, often causing up to 100% crop failure during severe attacks on important crops such as cowpea (Fig. 1). Many decades of conventional breeding efforts have failed to control this pest. However, recent research in U.S. universities and at the International Institute of Tropical Agriculture based in Ibadan, Nigeria, shows that this pest can be controlled by applying biotechnology tools.

This is just one of the myriad problems facing food production systems in Africa for which biotechnology can provide at least some solutions. Although biotechnology has potential downsides, the major "concerns" in Africa are not so much about justifying its role in agricultural productionthe "why" question. It is conceivable that the millions of dollars being wasted each year by antibiotech activists elsewhere could go a long way to help build badly needed capacity for agbiotech research in Africa! The key issues revolve around questions of where, when, how, and who will do biotechnology for Africa's benefit? If we are thinking of ultimate answers, then there is probably only one answer: biotechnology for Africa should mostly be done in Africa and mostly by Africans themselves, now. And yes, this is being realistic, and it can be done, if there is consensus and goodwill.

Despite many years of agricultural and other "development" aid and promises by different agencies related to increased food security and poverty eradication, those of us who live in Africa do not have confidence that things are getting any better. Because of this history, some are either pessimistic or skeptical, but the majority remain cautious and optimistic, that modern biotechnology opens new opportunities to address constraints that have led to declining harvests in farmers' fields in the midst of an expanding population. Richard Manning (2000) makes a good point when he suggests that one way to feed the increasing world population is to help "third world scientists to feed their own people, while ensuring sensitivity to culture and environment that we missed in the first green revolution."

For SSA, the pertinent question is, how does the international community of public and private institutions and donors, governments, scientists, and other actors help African scientists (and farmers!) to feed their own people? It is crucial that scientific information reaches farmers in the rural areas who have space to practice farming and that other actors such as agricultural scientists and extensionists interact with farmers to attain acceptance and use of new technologies for sustainable food production and development. In this regard, we must have it in mind that life science technologies that offer hope to farmers, such as agbiotech, belong to the farmer. We must also ensure that the technology not only reaches farmers but that they understand it and are empowered to use it. Furthermore, our starting point is not the "ignorant peasant" but the practices, techniques, experience, and knowledge of the African farmer built over the centuries (Duprez and DeLeener, 1988).

A good example of how biotechnology can reach rural farmers involves a special program by the Biotechnology Development Co-operation of the Netherlands Government, the Kenyan Ministry of Research, Science and Technology, and the small-scale farming system stakeholders. The program structure is designed to ensure that biotechnology reaches the small farmer (end-user) through a bottom-up approach steered by the Kenya Agricultural Biotechnology Platform. The composition of farmers includes male and female farmers, oxen owners, different age groups from different subvillages, etc. Projects under the Kenya Agricultural Biotechnology Platform funding bring together collaborators who include scientists from research institutions such as universities, national agricultural research centers, and farmers. A Farming Systems Research Program ensures that farmers participate in the research as partners with scientists, extensionists, and other actors and enables scientists also to utilize indigenous knowledge in research and development.

This prevents "cut and paste" approaches that may be foreign market-driven and which tend to provide short-term, quick-fix solutions to unique problems faced by small-scale farmers in Africa who have developed their own unique crops, cropping, and farming systems that cannot be changed without their full and careful involvement. Since 1992, Farmers Research Groups and Farmers Extension Groups, established along the lines of Farming Systems Research Programs, have been in existence in the Lake Zone of Tanzania for purposes of farmer participatory research. This experience shows that such participatory methods increase farmers' inputs in the decision-making process as well as in the dissemination of research products through their involvement in field trials, farmers' and "on-station" field days, PRA surveys, and farmer-to-farmer diffusion of information through Village Extension Workers rather than institutional extension.

Since Farmers Research Groups represent different geographic zones and hence different agro-ecological and farming systems, linkage mechanisms that bring together their experiences need to be established to allow horizontal and vertical dissemination of technologies as well as collaboration in the SSA region. Obviously, this is not the only way that research results from the laboratory reach farmers' fields, but it illustrates the fact that applied agbiotech research can similarly be targeted and tied to meet specific needs of rural farmers, both in the short-and long-term, in the face of scant resources. With African farmers and scientists working together to set the research agenda, there is hope that the research will focus on uniquely African ("orphan") crops such as millet and sorghum that are very important in marginal, famine-prone regions such as the Sahel and Horn of Africa. Root and tuber crops such as yam, sweet potato, and cassava may also begin to receive the attention they deserve.

Although Africa lags far behind other regions when it comes to public information and awareness of biotechnology issues, excellent work is being done by organizations such as the Nairobi-based African Biotechnology Stakeholders Forum and South African-based AfricaBIO to educate the general public in biotechnology. Opportunities abound for scientists in Africa to get involved in these efforts that are urgently needed if Africans are going to decide for themselves what biotechnology can do for them rather than let others decide for them, especially anti-genetically modified organism activists! There is also urgency to educate policy makers in African governments and the private sector concerning the need to support and invest in biotechnology Research and Development (R&D).

At the same time, the international donor community needs to begin to trust Africans and allow them to manage their research agenda for themselves. They can take the cue from very successful initiatives undertaken by the Rockefeller Foundation in Africa. There are enough African scientists around to make a difference on farmers fields if resources are properly channeled for agricultural R&D. African scientists and science managers in government and other institutions as well as farmers, on the other hand, need to be efficient and faithful in the way they manage research programs and funds if they are going to be trusted with money by national and international donors. The current success in tissue culture-aided production and multiplication of disease-free planting materials for cassava, yam, banana, plantain, citrus, and flowers in countries such as Kenya and Ghana is attracting private sector companies who are seeing the potential to invest in successful new biotechnologies.

On November 8-11, 2000, the Strategic Alliance for Biotechnology Research in African Development (SABRAD) held a workshop in Accra, Ghana, that brought together more than 150 participants from southern, East, Central, and West Africa as well as partners from the U.S. 1890 Land Grant Universities, U.S. Department of Agriculture, Food and Agricultural Organization of the United Nations, United Nations Environment Program, International Agricultural Research Centers, other non-governmental organizations, private companies, and journalists. International Agricultural Research Centers were represented by the Mexican-based International Maize and Wheat Improvement Centre and International Institute of Tropical Agriculture.

The theme of this first SABRAD Workshop was "Enabling Biotechnology for African Agriculture." Increasing education and awareness and formulation of regulatory (policy) frameworks that would allow access to modern biotechnology for R&D were identified as key priorities for enabling biotechnology for African development that targets resource-poor rural farmers. The one thing that was unique at the Accra meeting was that Africans themselves were at the center of discussions to work out plans for enabling biotechnology to take root in their respective countries. The action plans agreed upon will be implemented through networking between regions. The ultimate socio-economic impact is food self-sufficiency and improved living conditions of resource-poor farmers who were identified as the target recipients for products generated from biotechnology applications.

We live in a world that has become an increasingly interdependent "global village" due to advances in information and transportation technology. In this global village, millions have plenty of food to throw away, while millions of others die daily because they have nothing to eat. It is not always true that those with surplus food do not care about those who die in near and far away places! In Africa itself, there are many that have plenty of food, acquired either genuinely or by stealing public wealth, and who still watch their hungry neighbors die helplessly. Although Africans are thankful for development and relief aid, they are uncomfortable about their condition of continuous dependence on handouts that come in many forms, including food and expatriate foreign aid, with no permanent solutions apparently in sight. The SABRAD initiative is one step in the right direction that deserves support from all those who want to help African scientists and farmers to feed their own people.

AGRICULTURAL BIOTECHNOLOGY: UPDATED BENEFIT ESTIMATES

U.S. farmers have adopted genetically modified crop varieties rapidly since their introduction. By 2000, roughly one fifth of U.S. corn acreage, over half of the soybean acreage, and almost three-quarters of the cotton acreage was planted to crops genetically modified to be resistant to insects and/or herbicides.

This report updates our previous estimates of the benefits associated with the adoption of genetically modified crop varieties.

Updated impact estimates have been calculated for corn, cotton and soybeans. A discussion of potential impacts for potatoes is included. Each of these crops delivers a unique set of benefits to growers who adopt them. These benefits largely depend on pest control issues particular to each crop and whether other effective and affordable pest control options are available. Understanding the reasons why growers are adopting these varieties is critical in evaluating the their impact.

Insect resistant Bt corn varieties have allowed farmers to control the European Corn Borer, an insect pest that is difficult to control using conventional insecticides because it tunnels into the corn stalk soon after hatching, and is protected from the insecticides. Prior to the introduction of Bt corn, few growers were spraying for the corn borer.

Instead, yield losses sometimes reached 300 million bushels of corn per year. With Bt corn, losses to the corn borer are eliminated. The primary benefit of Bt corn varieties has been increased yields. In 1999, it is estimated that 66 million bushels of corn were saved from the corn borer, or the equivalent of production on nearly 500,000 acres. Modest reductions in corn insecticide use have also been observed since the introduction of Bt corn.

Cotton growers have adopted genetically modified varieties faster than growers of any other crop. Both insect and herbicide resistant varieties have been adopted widely. Insect protected varieties of Bt cotton provide control of three of the most destructive insect pests in cotton: tobacco budworm, cotton bollworm and pink bollworm. Growers were managing these pests with several insecticide treatments per year. In some areas, the efficacy of these insecticides was diminished by the development of insecticide resistant insect populations.

Bt cotton has allowed growers to reduce insecticide use and attain better control of these pests, which has resulted in increased yields. It is estimated that cotton growers reduced insecticide use by 2.7 million lbs and made 15 million fewer insecticide applications per year due to the introduction of Bt cotton. Cotton production has also increased, by 260 million lbs per year. Net revenues are estimated to have increased by $99 million in 1999.

Herbicide resistant cotton varieties provide growers with effective weed control programs that eliminate some of the problems associated with conventional programs. Until 1995, cotton growers did not have any broadleaf herbicides that could be used over a growing cotton crop without causing crop injury. Instead, herbicide treatments on small cotton required time consuming treatments using special equipment to treat weeds between the rows without contacting the crop. With the introduction of herbicide resistant cotton, growers can use broad spectrum herbicides over the growing crop with minimal crop injury. The introduction of herbicide resistant cotton varieties has led to a reduction in the number of herbicide applications made by cotton growers, by 19 million in 2000.

Insect and virus resistant potato varieties have been introduced that have great potential to decrease insecticide use in that crop. However, the recent introduction of a highly effective conventional insecticide and the refusal of processors to accept genetically modified potatoes have limited the adoption of these new varieties. Finally, herbicide tolerant soybeans offer growers effective weed control with a simple, flexible program that has allowed many to reduce weed control costs. Prior to the introduction of herbicide tolerant soybeans, growers chose from many herbicides, often applying three or more active ingredients, some of which would cause damage to the growing soybean plants, or cause harm to corn crops that commonly follow soybeans. Herbicide tolerant soybeans allow growers to rely on one herbicide to control a broad spectrum of weeds without harming the current or rotation crops. The primary benefit of herbicide tolerant soybean varieties has been a reduction in weed control costs, of $216 million per year in 1999. Growers have also reduced the number of herbicide applications, by 19 million in 1999.

U.S. farmers have adopted genetically modified crop varieties rapidly since their introduction in the mid-1990's. Genetically modified varieties of three major crops including corn, cotton and soybeans have been planted in the United States. (See Table 1.) A genetically modified potato variety has also been planted on limited acreage. Each of these crops delivers a unique set of benefits to growers who adopt them. These benefits largely depend on pest control issues particular to each crop and whether other effective and affordable pest control options are available. Understanding the reasons why growers are adopting these varieties is critical in evaluating the impact these technologies have had on US agriculture.

Previous estimates of aggregate benefits to U.S. farmers indicate higher yields, lower costs and ease of management. In addition, pesticide use has declined since the introduction of genetically modified varieties (Gianessi, et al. 1999, 2000; Carpenter, et al.). This report updates previous estimates of aggregate benefits resulting from farmers' adoption of genetically modified varieties for corn, cotton, potatoes and soybeans. The aggregate impacts of the introduction of genetically modified crop varieties are summarized in tables 2-5. These tables present aggregate estimates of changes in net revenue, increases in production, decreases in pesticide use and decreases in pesticide acre treatments.

An application acre is the number of different active ingredients applied per acre times the number of repeat applications, and differs from the number of trips over the field, as one trip across the field to apply two active ingredients is treated as two applications, as is two treatments each containing a single ingredient. Bt corn insecticide reductions in number of acres treated HT cotton-herbicide tolerant cotton, including BXN and Roundup Ready varieties.

CORN

Bt field corn varieties were introduced for planting by U.S. farmers in 1996, offering built-in insect control. Adoption of Bt corn varieties was swift, rising to 26% of corn acres planted by 1999, the fourth year on the market. Adoption decreased to 19% in 2000, which is believed to be due largely to historically low target pest pressure in 1998 and 1999. Figure 1 shows adoption of Bt corn since its introduction in 1996. Corn is the largest acreage crop grown in the U.S. Planted acreage totaled 80 million in 2000, which represents approximately one-quarter of the acreage planted to all crops in the U.S. Average corn yields totaled 134 bushels per acre in 1998 and 1999 (USDA NASS 2000a). Corn for grain production was estimated at 9 billion bushels in 1999 (USDA

NASS 2000c) with a total value of production of $18 billion. This represents approximately 20% of the value of all crops grown in the U.S. (USDA NASS 2000d). All 48 coterminous states have corn acreage, and, in many states, corn is the single most important crop in terms of acreage and production value. Corn production is concentrated in the Midwest, where ten states account for 85% of U.S. acreage and production. Individually, the states of Illinois and Iowa account for over 10 million acres of corn each (USDA NASS 2000a).

The major use of corn produced in the U.S. is as a livestock and poultry feed, accounting for 60% of use, while food, seed and industrial uses (including sweeteners, fuel alcohols and starch) account for approximately 19% of use. Exports account for the remaining 21% of use (USDA NASS 2000b). The U.S. is by far the largest corn producing country, growing 40% of the world's total production. China is the second largest producer, growing 21% of the world's total. Brazil, Mexico, France and Argentina are also major corn producing countries. The U.S. dominates the corn export market, accounting for 75% of the world's total exports. Major export markets for U.S.-grown corn are Japan (29%), Korea (12%), Mexico (10%) and Taiwan (8%). Western Europe accounts for less than 1% of U.S. corn exports (USDA NASS 2000b).

Bt corn varieties were genetically modified to express an insecticidal protein from the soil bacterium *Bacillus thuringiensis* (Bt). The Bt protein is selectively active against lepidopteran, or caterpillar insects, including the European Corn Borer (ECB), a major insect pest of field corn in the U.S. The ECB ingests the Bt protein when it starts to feed on the plant. The toxin binds to the gut membranes, which causes the ECB larvae to die. Due to the difficulty in scouting for this pest and the importance of timing insecticide application before the caterpillar bores into the corn stalk and is protected from insecticides, it is estimated that less than 5 percent of field corn acreage in the U.S. "Corn Belt" was being treated with insecticides for the ECB prior to the introduction of Bt corn varieties.

Researchers have sought alternative methods to control the corn borer, although none has proven effective on a wide scale. Traditional breeding efforts to select for natural resistance to the corn borer resulted in the development of varieties with intermediate levels of resistance that were widely used into the mid 1970s. However, acreage planted to these varieties decreased dramatically due to the introduction of much higher-yielding susceptible hybrids, which could produce higher yields than the resistant hybrids, even after sustaining high levels of damage from the corn borer. Another extensive research program was undertaken by the U.S. Department of Agriculture (USDA) to identify natural predators of the corn borer. The program resulted in the introduction of 24 species of parasites in U.S. corn production areas, of which only six became established. While these beneficial insects have provided control of the corn borer in some years in some areas, their impact has been limited.

Several different types of Bt field corn have been registered. The U.S. Environmental Protection Agency (EPA) has registered five unique Bt corn products: 176 (Novartis Seeds and Mycogen Seeds), Bt11 (Northrup King/Novartis Seeds), Mon 810 (Monsanto), DBT418 (Dekalb Genetics), and CBH 351 (Aventis, was AgrEvo). The Mon 810 and Bt11 events are used in production of YieldGard corn. Event 176 is sold as KnockOut by Novartis and Nature Gard by Mycogen. The DBT418 event is sold as BT-Xtra. The CBH351 event is sold as Starlink by Aventis. These events vary in the specific Bt protein that is produced, how much of the protein is produced, and where it is produced in the plant, thus affecting ECB control. The Starlink corn hybrid contains the Cry9C protein while all the other Bt hybrids contain the Cry1A(b) or Cry1A(c) protein. The Cry9C protein binds to a different site in the insect's gut. Cry9C is also toxic to black cutworms. The registration for Starlink varieties was voluntarily withdrawn by Aventis in October 2000. Starlink varieties had been planted on less than 1% of total U.S. corn acreage.

The impacts of Bt corn include increased yields and reduced pesticide use. However, the main advantage of Bt field corn has been increased yields, as only a small proportion of U.S. field corn acreage was sprayed for the ECB prior to the introduction of Bt varieties, as noted above. Yield losses due to the corn borer vary from year to year with infestation levels, which are generally unpredictable from one year to the next. Figure 2 shows ECB infestation levels in Illinois for 1943 to 1999. Largely uncontrolled until the introduction of Bt corn varieties, the ECB has caused production losses that have ranged from 33 million bushels to over 300 million bushels per year (USDA APHIS). Bt corn varieties have been shown to provide a very high level of protection from the corn borer. Plants sustain only minute damage as the corn borer larvae attempt to feed.

The benefits that growers realize from planting Bt corn varieties depend on the level of infestation. In light infestation years the benefits may not be great, while in heavy infestation years growers will realize substantial yield increases. An average of available research results comparing yields from Bt and non-Bt corn fields indicates that growers experienced a yield advantage of approximately 12 bushels an acre in 1997, 4.2 bushels an acre in 1998 and 3.3 bushels an acre in 1999. The seed price premium, or technology fee, for Bt corn was approximately $10 an acre in 1997 and 1998 and $8 in 1999. Assuming corn prices of $2.43 a bushel in 1997, $1.95 a bushel in 1998 and $1.90 a bushel in 1999, the average income changes for Bt corn were an increase of $18 an acre in 1997, a decrease of $1.81 an acre in 1998 and a decrease of $1.73 an acre in 1999. It is expected that in 10 of the 13 years from 1986 to 1998, corn borer infestations in the Corn Belt were such that corn growers would have realized a gain from planting Bt corn (Monsanto). Table 6 shows aggregate cost and benefits estimates of Bt corn for 1997 through 1999.

The greatest impact of Bt corn varieties is the increase in production it provides through reduced yield losses due to the ECB. Table 7 shows the increase in production due to reduced yield losses due to Bt corn for 1997 through 1999. In 1999, growers were able to reduce their yield losses by 66 million bushels, the equivalent to production on nearly 500,000 acres of corn production that would otherwise have been destroyed by the corn borer.

Reductions in insecticide use are also expected due to the introduction of Bt corn varieties, though these reductions are anticipated to be modest due to low levels of insecticide use for ECB prior to the introduction of Bt corn. Attributing any observed changes in insecticide since 1995 to the introduction of Bt corn is necessarily problematic for several reasons. First, insecticides are typically used for control of several target pests. USDA pesticide use surveys do not report insecticide use by target pest, which means that isolating insecticide use targeted solely to the ECB is impossible. Second, insect populations by their nature are highly variable from year to year, which makes trends difficult to discern. Finally, the introduction of new products and development of resistance in insect populations to older products also drive shifts in insecticide use. Five insecticides are currently recommended for control of ECB: Bt (foliar spray), chlorpyrifos, permethrin, lambdacyhalothrin, and methyl parathion. With the exception of foliarly applied Bt products, these insecticides are typically used for several target pests, including cutworms, rootworms, armyworms as well as the ECB.

Comparing 1995, the year before Bt corn varieties were introduced, to 1999, the use of these five insecticides declined. (See Table 8.) Of interest is the decline in use of four of the insecticides between 1995 and 1999: chlorpyrifos (-2%), permethrin (-1%), Bt (-1%), and methyl parathion (-2%). The decrease in the use of these insecticides may be due to the introduction of Bt corn varieties and the resulting reduced need for sprays targeted at the ECB. However, several other explanations are also possible to explain the change from 1995 to 1999. Lambdacyhalothrin was introduced in 1996, primarily for the treatment of cutworm in-season, which has likely displaced some of the use of the other insecticides, such as chlorpyrifos. Adult corn rootworm beetle populations have reportedly developed resistance to methyl parathion in some areas, which may account for part of the decline in its use. Finally, these reductions may be due to a decline in insecticide treatments targeted at pests other than the ECB. It is likely that all of these factors have affected use patterns of these insecticides.

The aggregate reduction of the percent acreage treated with the five insecticides is 6%. Assuming that 3% of the change is attributed to the introduction of lambdacyhalothrin, implies that a 3% decline occurred as a result of changes in the target pest complex, including ECB and other pests. For analytical purposes, it is assumed that one-half of the decline in the usage of the four insecticides (1.5%) was due to the introduction of Bt corn, implying over 1 million fewer acres sprayed for ECB.

COTTON

In 2000, 72% of U.S. cotton acreage was planted to insect and/or herbicide resistant varieties: 39% *Bt*, 61% herbicide resistant and 28% "stacked" varieties with both insect and herbicide resistance (USDA AMS). Figure 3 and Table 9 show adoption of genetically modified cotton varieties since 1996.

Cotton is the fifth largest crop in the U.S., in terms of acreage in production, planted on over 15 million acres in 2000 (USDA NASS 2000a). In 1999, nearly 17 million bales of cotton were harvested, worth nearly $4 billion (USDA NASS 2000d). Cotton production is concentrated in the south and west. Texas is the largest cotton producing state, accounting for 40% of total U.S. acreage and 30% of total production in 2000. Georgia has recently expanded acreage to become the second largest cotton producing state, in terms of area in production, planting 9% of U.S. total acreage in 2000. California, North Carolina and Arizona each have nearly 1 million acres in cotton production (USDA NASS 2000a).

Bt Cotton

Bt cotton varieties were introduced in 1996, providing control of three major cotton insect pests: tobacco budworm, cotton bollworm and pink bollworm. Cotton bollworm and tobacco budworm are prevalent in the Southeast and Mid-South production areas, while pink bollworm infests cotton acreage primarily in western states such as Arizona and California.

These varieties offer an alternative to conventional insecticide spray programs. Insecticides were used on 75% of the total cotton acreage before the introduction of *Bt* varieties (USDA NASS). In 1995, the year before *Bt* varieties were introduced, it was estimated that 2.4 insecticide applications per year, on average were made to control bollworm/budworm across all cotton producing states, and that a 4% yield loss was incurred due to these two pests. Tobacco budworm infestations were particularly heavy in 1995, causing severe yield losses in some areas.

The worst damage was sustained by Alabama growers, who on average experienced a 29% yield loss due to bollworm/budworm (Williams 1996). These losses were attributed to the ineffectiveness of pyrethroid insecticides against budworm, due to the development of resistant populations in some states.

Two surveys provide estimates of Bt cotton adoption. The USDA Agricultural Marketing Service estimates are based on informal surveys of cotton ginners, seed dealers, extension agents and others. In addition, the National Cotton Council sponsors an annual survey that provides estimates of adoption of Bt cotton varieties.

The Cotton Council survey is coordinated by entomologists in each of the states, collecting information derived from surveys of county agents, extension specialists, private consultants and research entomologists. Tables 10 and 11 show estimates of adoption of Bt cotton from USDA AMS and the Cotton Council. Adoption estimates for 2000 are not yet available from the Cotton Council.

The adoption of *Bt* varieties was extremely rapid in some areas and has been slower in others. After a year of very high budworm populations and damage in 1995, growers in Alabama adopted the new technology at an extremely rapid rate, planting over 60% of total acreage to *Bt* varieties in 1996 (Williams 1997; USDA AMS). In 2000, 65% of cotton acreage in Alabama was in *Bt* varieties (USDA AMS). Florida and Mississippi also adopted the technology on over 30% of cotton acreage in 1996. By 2000, four states had adoption rates of over 75%: Florida, Louisiana, Mississippi and Tennessee (USDA AMS). Figure 4 shows adoption of *Bt* varieties by state for 2000.

Two major cotton producing states have had very low adoption rates thus far, which skews the national adoption rate. Texas, which accounted for 40% of cotton acreage in 2000, has only adopted *Bt* cotton on a small scale, at only 10% in 2000.

Adoption has been hindered by the lack of stripper varieties appropriate for growing conditions in Texas. California also has low adoption rates, which is at least partially due to a unique law in that state that controls which varieties may be planted. Demand for *Bt* cotton varieties in California is expected to be relatively low, however, since most producing areas of California are not infested with the three target pests of *Bt* varieties.

Table : USDA AMS Bt Cotton Adoption Estimates

	percent of planted acres				
	1996	**1997**	**1998**	**1999**	**2000**
Alabama	65	80	61	75	65
Arizona	11	64	57	56	57
Arkansas	15	14	14	22	60
California	0	2	5	8	6
Florida	42	79	80	66	75
Georgia	23	34	47	51	48
Louisiana	21	32	71	67	81
Mississippi	37	45	60	66	75
Missouri	0	0	0	1	5
New Mexico	9	13	38	28	36
North Carolina	3	2	4	37	41
Oklahoma	2	7	2	47	55
South Carolina	16	27	17	84	71
Tennessee	0	6	7	59	76
Texas	1	4	7	9	10
Virginia	0	3	1	16	41
US	12	18	23	32	39

Source: USDA AMS

Table : Cotton Council Bt Cotton Adoption Estimates

	percent of planted acres			
	1996	1997	1998	1999
Alabama	77	68	63	66
Arizona	21	61	71	63
Arkansas	14	13	12	16
California	0	0	1	3
Florida	43	60	50	56
Georgia	30	38	44	58
Louisiana	15	38	60	61
Mississippi	35	43	58	67
Missouri	0	1	1	2
New Mexico	1	6	35	23

Contd...

	percent of planted acres			
	1996	**1997**	**1998**	**1999**
North Carolina	6	3	12	20
Oklahoma	6	8	8	21
South Carolina	14	31	39	81
Tennessee	1	3	19	68
Texas	2	5	5	8
Virginia	0	0	3	7
US	13	17	21	29

Source: Williams

The impact of the adoption of Bt cotton varieties has been a reduction in yield losses due to Bt-target pests, reductions in insecticide use, and cost savings. For Alabama growers, yield losses due to bollworm/budworm were drastically reduced from levels experienced in 1995. In Central Alabama, where the average yield loss due to bollworm/budworm was 55% in 1995, losses varied between 2 and 7% from 1996 to 1999. The number of insecticide applications per year was reduced from 10 to between 0 and 2 (Williams). Yield losses due to Bt target pests have declined in 12 of 16 reporting states. Table 12 shows percent yield losses due to cotton bollworm, tobacco budworm and pink bollworm by state for 1985 to 1999.

The major impact of the adoption of *Bt* varieties is a reduction in insecticide use. There are two sources of cotton insecticide use data available. The first is the Cotton Council survey, which estimates insecticide use by target pest. The number of insecticide treatments for cotton bollworm, tobacco budworm and pink bollworm decreased in 1999 compared to 1995 in 12 of 16 reporting states. Table 13 shows the number of insecticide applications directed toward cotton bollworm, tobacco budworm and pink bollworm by state from 1986 to 1999.

As part of the Cotton Council survey, three states report insecticide use separately for Bt acreage and conventional acreage. (See Table 14.) In all three states, the number of insecticide treatments directed at Bt target pests was lower on Bt acreage. However, treatments for other insects (except for boll weevil) were higher on Bt acreage. This is believed to be due to the emerging importance of pests that were previously considered secondary pests, which were previously controlled with treatments at Bt target pests. However, the reduction in treatments for Bt targets is larger than the increased number of treatments towards these secondary pests in all three instances. Finally, in Louisiana and Tennessee, the number of treatments for boll weevil is higher on Bt acreage than on conventional acreage. This is likely due to higher adoption rates for Bt cotton in boll weevil eradication areas.

USDA pesticide use data record insecticide use by active ingredient without indication of target pest. Comparing USDA pesticide use data for 6 states for 1995 to 1998 and 1999 shows a dramatic reduction in the use of insecticides used to control cotton bollworm, tobacco budworm and pink bollworm. Table 15 shows reductions in insecticide use for those insecticides that are recommended for cotton bollworm, tobacco budworm and pink bollworm control. This difference in use was adjusted to account for reduction in planted acreage since 1995. The use of cotton insecticides for Bt target pests declined by over 2 million pounds between 1995, the year before Bt varieties were introduced, and 1998, and by 2.7 million pounds by 1999. These reductions account for 10% and 14% of all insecticides used in those six states in 1995. The number of insecticide applications has also declined, by 8.7 million applications in 1998 and by 15 million applications in 1999, or 13% and 22% of the total number of insecticide applications in those states in 1995.

Several surveys have found that growers are achieving higher yields and attaining higher profits by planting *Bt* varieties, due to better pest control and decreased insect control costs. The average increase in net returns from 5 studies in 7 states including 91 comparisons in 1999, comparing *Bt* to conventional varieties was $20.81/acre, taking into account the technology fee. (See Table 16.) On average, per acre insect control costs were $14.28 higher. This increased cost was outweighed by a yield increase of 9%, or 55 lbs/acre. These yield and revenue impacts, if realized over all 4.67 million acres of *Bt* cotton in 1999, would result in a $99 million increase in revenues and 260 million lbs increased cotton production.

Herbicide Resistant Cotton

In 1995, BXN cotton was introduced. BXN cotton is tolerant to bromoxynil (Buctril), a post-emergence broadleaf herbicide already registered for use in corn and small grains. Roundup Ready cotton varieties became available in 1997, presenting growers with another option for post-emergence weed control using glyphosate (Roundup).

Nearly all U.S. cotton acreage is currently treated with herbicides. In 1998, 95% of U.S. cotton acreage was treated with herbicides (USDA NASS). The average number of herbicide applications per treated acre was 2.6 in 1995, while 34% of the acreage received 3 or more applications. An average of 2.7 different active ingredients per acre were used in 1995, with 24% of the treated acreage receiving 4 or more active ingredients (USDA ERS 1997). Primary weed species are morningglory (*Ipomoea* spp.), pigweed (*Amaranthus* spp.), nutsedge (*Cyperus* spp.), cocklebur (*Xanthium* spp.) and johnsongrass (*Sorghum* spp.) (Byrd).

Prior to 1995, cotton growers did not have any broadleaf herbicides that could be used over the top of a growing cotton crop without the potential to cause crop injury. Instead, growers would use directed, post-emergence applications of nonselective herbicides, using specialized equipment to avoid herbicide contact with the growing crop plants, and cultivation. Directed, post-emergence treatments require weeds to be shorter than the cotton crop. This height differential is sometimes difficult to achieve. Post-directed treatments on small cotton require time-consuming treatments and can damage plants if herbicides contact the plant foliage.

In 1996, a new post-emergence broadleaf herbicide became available for use over the top of growing cotton without causing crop injury. Pyrithiobac (Staple) is a selective broadleaf herbicide that may be applied post-emergence at any stage of crop growth. The first year it was available, it was used on 10% of U.S. cotton acreage, increasing to 23% in 1997 (USDA NASS). By 1999, its use had dropped back to 14% (USDA NASS), likely due to competition with weed control programs using newly introduced herbicide tolerant cotton varieties.

Table : Yield Losses Due to Tobacco Budworm, Cotton Bollworm and Pink Bollworm

	1985	1986	1987	1988	1989	1990	1991	1992	1993	1994	1995	1996	1997	1998	1999
															percent
Alabama	2.6	4.5	4.5	2.6	3.4	1.4	1.7	2.5	6.8	6.1 29.1	3.1	3.2	4.7	2.0	
Arizona	3.6	2.9	2.3	1.4	1.5	4.6	1.1	0.7	0.1	3.9	1.2	2.8	2.7	0.0	1.0
Arkansas	1.8	1.8	1.1	1.4	1.0	1.3	1.5	2.6	1.5	1.8	3.6	3.1	2.7	4.7	1.3
California	0.8	0.0	0.0	1.0	0.0	0.4	0.0	0.0	0.0	0.0	0.0	0.0	0.0	0.0	0.0
Florida	4.3	5.5	7.2	4.4	5.5	9.1	4.0	3.0	3.0	3.7	3.9	3.3	4.3	2.1	0.7
Georgia	3.6	4.1	2.1	2.7	4.7	5.0	1.0	1.8	1.9	1.9	2.8	1.9	2.5	2.8	0.8
Louisiana	4.0	5.3	2.7	2.3	5.8	4.0	3.5	7.5	3.6	2.9	3.2	2.3	1.9	1.5	0.8
Mississippi	1.3	4.9	4.6	4.9	4.6	4.5	0.6	3.9	3.9	4.1	8.0	1.9	2.4	4.2	2.5
Missouri	1.0	0.6	1.9	7.5	0.0	1.1	1.6	1.7	0.9	0.5	1.4	1.3	1.3	6.1	0.7
New Mexico	4.5	3.9	0.8	2.4	2.5	5.0	3.4	1.6	3.0	1.6	3.3	10.6	8.6	6.1	3.7
North Carolina	8.5	2.5	0.9	5.2	8.6	3.6	12.6	4.3	2.3	6.4	3.7	5.2	5.3	4.4	4.1
Oklahoma	5.0	1.8	1.9	1.3	0.5	1.2	1.6	3.3	3.1	1.7	2.5	3.3	3.0	4.0	5.0
South Carolina	4.3	3.2	2.1	4.6	2.6	2.4	1.9	2.8	3.4	3.4	4.7	5.0	4.7	3.5	1.9
Tennessee	3.0	2.0	2.1	1.6	0.8	0.3	1.0	1.0	6.7	0.5 10.8	0.8	1.0	5.9	1.6	
Texas	3.8	2.3	1.9	0.9	1.5	1.1	1.8	1.4	0.5	0.8	1.0	3.0	1.7	1.3	0.4
Virginia	15.0	10.0	2.4	3.4	4.7	12.5	4.0	0.2	0.0	0.5	1.0	2.0	0.3	0.0	0.3

Source: Williams

Table : Insecticide Treatments for Tobacco Budworm, Cotton Bollworm and Pink Bollworm

	1986	1987	1988	1989	1990	1991	1992	1993	1994	1995	1996	1997	1998	1999
	number of applications per year													
Alabama	5.2	4.9	2.0	3.7	2.5	2.9	2.9	4.8	4.4	6.7	0.1	0.5	1.4	0.4
Arizona	2.2	6.3	2.9	3.9	7.4	3.2	1.1	0.1	2.9	2.9	2.0	1.1	0.1	0.4
Arkansas	2.9	1.3	2.9	2.6	2.4	3.3	3.7	3.6	3.0	4.6	2.3	1.6	3.1	0.5
California	0.1	0.0	0.6	0.2	0.1	0.3	0.0	0.0	0.0	0.0	0.0	0.0	0.0	0.0
Florida	9.0	7.1	5.6	8.4	7.2	5.0	5.0	5.3	5.3	5.7	1.1	1.0	2.0	0.5
Georgia	7.3	6.2	4.4	4.3	5.0	4.9	3.4	2.7	4.3	3.4	1.7	2.5	1.5	0.6
Louisiana	5.5	4.3	3.7	5.8	5.0	3.5	5.8	4.7	4.8	4.7	3.9	3.2	3.5	1.2
Mississippi	4.6	4.7	3.7	4.8	3.4	1.5	5.1	4.3	4.1	5.7	2.2	2.5	2.5	1.3
Missouri	0.3	0.6	1.5	0.0	0.1	0.3	0.3	0.9	0.8	1.3	0.1	0.7	1.6	0.9
New Mexico	1.1	0.1	0.4	0.6	1.1	0.8	0.3	0.3	0.7	0.8	1.1	0.7	2.0	1.1
North Carolina	3.5	2.7	2.6	2.9	2.8	3.1	3.0	2.5	3.6	2.6	3.1	2.0	3.0	1.9
Oklahoma	1.8	1.1	1.3	0.8	0.9	1.3	1.1	2.3	0.7	1.8	1.7	1.9	1.5	0.7
South Carolina	3.1	2.8	3.7	3.5	3.8	3.8	3.7	4.9	4.4	4.7	4.2	3.3	3.4	1.1
Tennessee	0.4	0.5	0.3	0.4	0.3	0.6	0.1	2.0	0.3	2.9	0.2	0.3	2.7	0.6
Texas	0.8	1.2	0.8	0.7	0.7	1.1	0.7	0.4	0.7	0.5	0.7	0.6	0.6	0.1
Virginia	2.0	1.0	1.4	0.8	1.5	2.5	0.2	1.0	1.7	1.5	1.0	1.2	2.2	1.8

Source: Williams

The adoption of BXN and Roundup Ready varieties has been driven largely by the ease and convenience of avoiding early directed post-emergence herbicide applications, as well as having new tools to control particular weed problems. Along with the introduction of Staple, growers now have three new post-emergence herbicides for broadleaf weed control. This is especially valuable in areas where weeds have become resistant to other commonly used herbicides.

BXN cotton varieties were first planted in 1995, when EPA approved the use of bromoxynil for cotton. On May 18, 1995, EPA established time-limited tolerances for residues of bromoxynil on cottonseed, which allowed the use of bromoxynil over BXN cotton varieties. The accompanying registration allowed 200,000 acres of cotton to be treated with bromoxynil. On June 18, 1997, EPA renewed the time-limited tolerances, with an accompanying registration that increased the allowable treated acreage to 400,000 acres. On May 13, 1998, EPA established permanent tolerances for bromoxynil in cotton, with an accompanying registration allowing 1.3 million cotton acres to be treated with bromoxynil (US EPA 1998).

Bromoxynil controls many broadleaf weeds but does not control grasses, which makes continued use of soil applied herbicides likely. In particular, adoption has been high in some areas due to its effectiveness on morningglory and cocklebur. However, bromoxynil does not provide effective control of sicklepod, which limits adoption in areas where that weed is prevalent.

Adoption of BXN cotton varieties was very low until 1998. By 1998, adoption was almost 6%, rising to nearly 8% in 1999 and falling slightly to just over 7% in 2000. With current limitations on the total acreage that may be treated with bromoxynil, adoption can not increase substantially from current levels. The relatively small national adoption numbers hide what has been substantial adoption in some areas. In 1997, 7% of North Carolina and 4% of Tennessee acreage was planted to BXN varieties. Tennessee had adopted BXN varieties rapidly, 40% in 1998, but in 1999 reduced their acreage to 10% due to the commencement of boll weevil eradication and the demand for *Bt* varieties (Hayes). BXN varieties have not yet been stacked with the *Bt* trait. Missouri currently has the highest adoption rate for BXN varieties at nearly 40% of planted acres (USDA AMS). Table 17 shows adoption rates of BXN varieties by state since 1995.

Roundup Ready varieties, which are tolerant to the application of glyphosate (Roundup), became available in 1997. Roundup is a highly effective broad spectrum herbicide that controls both grasses and broadleaf weeds. Roundup Ready varieties are not perfectly tolerant to treatment with Roundup, however. Over the top treatment with Roundup may be made only until the 4-leaf growth stage, after which time, directed post-emergence treatments must be made.

Table : BXN Cotton Adoption

	1995	1996	1997	1998	1999	2000
				percent of planted acres		
Alabama						
Arizona	0.1	0.3	1.0	1.9	1.7	
Arkansas	0.3	0.5	2.0	27.8	41.3	21.9
California	3.0	8.7				
Florida	0.7	0.3				
Georgia	0.2	0.7	2.5	2.3	0.8	
Louisiana	0.2	0.3	0.6	8.0	8.7	3.4
Mississippi	0.3	0.2	3.1	6.2	15.0	11.0
Missouri	0.2	0.1	0.7	11.1	29.9	39.5
North Carolina	0.1	0.1	7.1	4.3	13.4	13.1
Oklahoma	1.6	0.7	2.9			
South Carolina	0.2	0.2	1.1	0.2	1.2	0.4
Tennessee	3.7	39.6	10.3	7.3		
Texas	0.1	0.2	0.4	1.1	1.8	
Virginia	3.4	11.2	16.4			
US	0.1	0.1	1.2	5.8	7.8	7.2

Source: USDA AMS

Since 1997, adoption of the technology has been rapid, growing from 4% in 1997 to 54% in 2000. Some areas have adopted the technology more rapidly than others. South Carolina planted 94% of its cotton acreage to Roundup Ready varieties in 1999 which is believed to be due to the effectiveness of Roundup on sicklepod, and DNA-resistant palmer amaranth (Murdock). Florida and Tennessee have also adopted at rates over 75% of planted acreage (USDA AMS). Table 18 shows adoption rates of Roundup Ready varieties by state since 1997.

Table : Roundup Ready Cotton Adoption

	1997	1998	1999	2000
				percent of planted acres
Alabama	1	13	40	60
Arizona	2	11	12	25
Arkansas	5	3	8	12
California	2	21		
Florida	7	54	83	
Georgia	2	35	55	66
Louisiana	1	3	12	54
Mississippi	11	10	27	53
Missouri	2	4	22	
New Mexico	1	2	2	4
North Carolina	10	32	58	65
Oklahoma	2	22	57	68
South Carolina	4	45	94	76
Tennessee	11	18	57	81
Texas	2	27	41	55
Virginia	3	19	39	49
US	4	21	37	54

Source: USDA AMS

Some growers may have realized cost savings by switching to Roundup Ready programs. While conventional programs may cost around $44 per acre, Roundup Ready programs cost between $23 and $47, including an $8/acre technology fee, depending on the number of applications and whether other soil-applied or post-emergence treatments are made. Staple is relatively expensive, with a program cost of approximately $57/acre. BXN programs cost approximately the same as a conventional program (Coble). In field trials of conventional programs with and without Staple compared to BXN and Roundup Ready programs, relative yields and net returns varied. Though these yield and return results are from few years and locations, at this point there appears to be no clear cut advantage of one program over another.

Adoption of BXN and Roundup Ready varieties has caused shifts in pesticide use patterns and reduced herbicide use in cotton production. Use of many of the most commonly used herbicides, such as trifluralin, fluometuron, and MSMA, has declined since 1994. Glyphosate use increased from use on 8% of planted acres in 1994 to use on 36% in 1999 (USDA NASS).

Table : Average Relative Yields for Conventional, BXN and Roundup Ready Programs for FL, GA, LA, MS, NC, and TN 1997-1998

Weed Control Program	Yield Relative to Conventional Program
Conventional	100%
Staple	95%
BXN-Buctril	93%
RR-conv.+1xRU	101%
RR-Treflan/RU	102%
RR-RU/Bladex+MSMA	93%
RR-RU as needed	96%

Sources: Brecke, Bridges, Hayes, Miller, Snipes, Wilcut

Herbicide use in cotton is expected to decline with the adoption of herbicide tolerant varieties. Application rates for conventional programs vary between 5.5 and 9 lbs./acre, compared to 2.75 to 4.5 lbs./acre for Roundup Ready systems and 2.8 to 4.45 lbs./acre for BXN systems (Coble). Indeed, USDA pesticide data shows a declining trend in herbicide use in cotton. USDA data show that average application rates decreased from 1.7 lbs/acre in 1997 to 1.63 lbs/acre in 1999. There has been a general decline in average cotton herbicide application rates since 1994. The total amount of herbicide applied to cotton acreage has also declined since 1995. Further, the number of applications made to cotton acreage has declined by 1.8 million applications in 1998 and 1.3 million applications in 1999 compared to 1994.

Table : Cotton Herbicide Application Rates 1994 to 1999

	lbs/acre	applications/acre
1994	2.02	2.96
1995	1.89	2.91
1996	1.57	2.53
1997	1.7	2.84
1998	1.61	2.76
1999	1.63	2.82

Source: USDA NASS

Much of the decline in total pounds of herbicides used in cotton since 1995 is likely due to the adoption of Staple and is probably not attributable to the introduction of herbicide tolerant varieties. Staple is used at a very low average rate, 0.05 lbs/acre, compared to other herbicides that are used at 0.09 lbs/acre to 1.18 lbs/acre (USDA NASS 2000). However, the decline in the number of applications is more likely due to the introduction of herbicide tolerant varieties and the associated use of broad spectrum herbicides, such as Roundup, which reduce the number of herbicides needed for effective weed control. Reductions have been adjusted to account for the decline in acreage in recent years.

The impact of the introduction of genetically modified cotton varieties has been a reduction in the total amount of insecticides used, reductions in the numbers of insecticide and herbicide applications, increased yields and increased returns to farmers. The use of insecticides declined by 2.7 million pounds by 1999, while the number of insecticide applications has declined by over 15 million applications. Herbicide use has also declined, although the decline in the total amount of herbicides applied may not all be attributable to the introduction of herbicide tolerant varieties. The number of herbicide applications has declined by over 1 million applications. Production increased by 260 million lbs per year. Returns have increased $99 million.

POTATOES

Bt potato varieties were introduced in 1996, and have been adopted on a very limited basis. One percent of the total U.S. potato acreage was planted to Bt potato varieties in 1996. By 2000, adoption was between 2 and 3 %. Three types of Bt potatoes are available: NewLeaf Bt potato, introduced in 1996; NewLeafPlus Bt potato with resistance to the potato leafroll virus, introduced in 1999; and NewLeafY Bt potato with resistance to potato virus Y, also introduced in 1999. Adoption figures reflect combined adoption of the three types of Bt potatoes.

The low adoption rates are due to a combination of factors. Early adoption of Bt potato varieties, prior to the introduction of stacked varieties, was hindered by the continuing need by potato growers to control other insect pests in addition to the Colorado Potato Beetle, which limited potential insect control cost savings. More recently, rejection by processors in response to McDonald's and other buyers' refusal to buy genetically modified potato products has limited adoption (Kilman). Finally, the introduction of a very effective conventional insecticide just prior to the introduction of Bt potatoes provided growers with an attractive alternative to Bt varieties.

Potatoes are grown on 1.4 million acres in the U.S., producing over 45 billion pounds of potatoes annually (USDA NASS 2000b). Total crop value was $2.8 billion in 1999 (USDA NASS 2000d). Fifty-three percent of annual U.S. potato production is located in the Pacific Northwest states of Idaho, Washington and Oregon. Other major potato producing states include Wisconsin, Colorado, North Dakota, Minnesota, Maine, California and Michigan (USDA NASS 2000c).

The majority of potatoes are processed into food products or ingredients. Sixty percent of total production is processed, of which 50% is made into frozen french fries, the rest is either dehydrated (19%), made into chips or shoestring potatoes (17%), other frozen products (9%), or other end products. Twenty-six percent of total potato production is grown for table stock. The remaining production is used either for seed (6%), is not marketed due to shrinkage or loss (7%) or is used for livestock feed or household use (USDA NASS 2000b).

Potatoes are annuals produced from certified potato seed pieces taken from the previous year's crop. Potatoes and seed potatoes are produced in the same manner, except that seed potatoes receive more intensive pest management to ensure disease-free planting stock (Schreiber, et al.).

The target insect pest of all Bt potato varieties is the Colorado Potato Beetle (CPB), which is more of a problem in the East than in Northwest production areas. CPB consumes plant foliage, which leads to reduced photosynthesis and consequent yield reductions. The CPB is the major defoliating insect pest of potatoes in the U.S. If uncontrolled, the CPB can completely defoliate all potato plants in a field by mid-season. Few insects have shown the potential to develop resistance to as broad a range of insecticides as the CPB. Eastern potato growers have found it necessary to change insecticides every few years as the CPB developed resistance to every insecticide used extensively against it (Casagrande). In the face of declining efficacy of conventional insecticide treatments, potato growers had resorted to other methods to control the CPB, such as using propane burners or constructing plastic-lined trenches in the borders around fields. Biological control of the CPB has been pursued using a fungal pathogen toxic to the beetle, but the success of this program was limited because fungicides used to control blights also killed the fungus. The introduction of predatory stink bugs has also been attempted, but difficulty attaining high enough populations of the stink bugs to provide control of the beetle limits the value of this practice (Ferro, et al.).

Aphids are another major insect pest of potatoes. Aphid infestation may reduce potato yields directly through feeding damage to the foliage, but causes more damage indirectly by transmission of viruses. The potato leaf roll virus (PLRV) is transmitted mainly by the green peach aphid and is prevalent in the Pacific Northwest. Aphids spread the virus when feeding, infecting expanding areas around the primary inoculum source as the aphids move from plant to plant. These expanding areas may engulf entire fields within a few weeks if the aphid populations are not controlled. Incidence of PLRV infection routinely approaches 100% in potato crops in the Northwest when insecticides are not used (Thomas, et al.). PLRV infection symptoms include leaf rolling, stunted plants, and discoloration of leaves, that become stiff, dry and leathery. PLRV infections reduce potato plant vigor and result in high yield losses (Hooker). PLRV also infects developing tubers, causing net necrosis, or stem end browning. Symptoms may not be evident at harvest, but may develop during storage. Losses in marketable yield can reach as high as 50-80% (Benttari).

The Russet Burbank potato variety is the dominant variety in the Northwest. The Russet Burbank is susceptible to potato leaf roll virus and the associated tuber net necrosis. The disease can limit production because the discolored tubers are not suitable for processing or table stock. The green peach aphid is the only important vector of the virus in the West and, because of this, is the key insect pest of Russet Burbank potato crops (Bishop, et al.).

Another major aphid-borne virus of potato is potato virus Y (PVY). PVY is transmitted mainly by the green peach aphid and the buckthorn aphid. Unlike PLRV, PVY is also transmittable mechanically by leaf contact and injury, although aphid transmission is the most important means of spread in the field. Diseases caused by PVY are severe mosaic, leaf-drop streak, and potato veinbanding mosaic. Reported losses range from 10 to 80% (Rowe). PVY does not cause price discounts due to damaged tubers as PLRV coes, but is considered one of the most damaging potato viruses in causing yield depression (Hooker). Control of PVY is especially important in the production of certified seed potatoes in Maine and Canada (Owens).

A new insecticide, imidacloprid, was commercialized for use in potatoes for the 1995 season and provides highly effective control of the Colorado potato beetle, aphids and other foliar-feeding insects. Imidacloprid is applied in furrow when the potato seed is planted, is taken up through the plant, persists in the plant and controls insects that feed on the plant. Following the introduction of imidacloprid in 1995, Eastern potato growers stopped using all of the cultural methods of beetle control (burners, trenches), except for crop rotation. USDA pesticide use data show use of imidacloprid on 34% of acres in 1999. Midwest and Eastern states treat a higher proportion of potato acreage with imidacloprid than in northwest production areas. Ninety-three percent of

Michigan potato acreage was treated with imidacloprid in 1999; 90% in Maine; and 81% in Pennsylvania.

The cost of an imidacloprid application at planting is $60 per acre (Michigan Potato Industry Commission). Imidacloprid has dramatically reduced populations of CPB in Eastern states. The reduction in CPB populations has been estimated to be as high as 99.9% in some locations. Thus, CPB populations throughout the Midwestern and Eastern potato growing regions of the U.S. have declined dramatically (Whalon, et al.). Yield losses due to CPB have become minimal to nonexistent (Michigan Potato Industry Commission).

With the exception of propargite, all insecticides commonly used in the Pacific Northwest are primarily to control CPB and aphids (Pacific Northwest Insect Control Handbook). In eastern production areas, all commonly used insecticides are directed primarily at CPB, except carbaryl, diazinon, dimethoate and methyl parathion (Cornell Cooperative Extension).

NewLeaf potatoes, with the Bt trait alone, were the first genetically engineered potato varieties that were made available to growers, in 1996. NewLeaf varieties only provide control of CPB. The best fit for these varieties would be in the Eastern states, where CPB is the primary insect pest and PLRV is not a concern. Surveyed growers who planted both NewLeaf and conventional Russet Burbank potatoes in 1998 made 1.35 fewer insecticide applications on their NewLeaf fields. In terms of total amount of insecticides used, NewLeaf fields required 0.48 lbs/acre less insecticide active ingredient. Eastern growers had the greatest reduction in the number of insecticide applications, from 3.12 to 1.77.

In the Pacific Northwest, growers who adopted the technology were those with a history of light infestation of pests other than CPB. The New Leaf potato technology fee was about $30 per acre in 1998. Avoiding an at-plant insecticide application cost of $60 per acre represented a saving of $30 per acre. If only one foliar application was needed during the year for aphids at a cost of $20 per acre, the grower could save $10 per acre. Some of the growers planting New Leaf were interested in supporting and trying the technology while others selected the New Leaf varieties for agronomic considerations. Most growers did not change their insect control practices and still used the at-planting systemic insecticide. The introduction of NewLeafPlus and NewLeafY potato varieties, with built-in virus protection in addition to protection against CPB, had the potential to provide growers with greater benefits in terms of reducing the need for insecticide treatments. NewLeafPlus and NewLeafY do not control the aphids that vector PLRV and PVY, but are protected from these viruses that the aphids transmit.

NewLeafPlus varieties are most valuable for growers in the Pacific Northwest, where PLRV is prevalent and destructive. In field trials conducted by NatureMark in 1998 and 1999, fewer insecticide applications were made to NewLeafPlus fields than to conventional Russet Burbank fields.

Table shows the average number of insecticide applications made to NewLeafPlus fields compared to conventional fields. NewLeafPlus potatoes are also expected to reduce losses due to net necrosis. In total, growers could save considerably in reduced losses due to net necrosis and in reduced insecticide costs. Including the $46 per acre technology fee for NewLeafPlus varieties, field trial results indicate an average savings of $85 per acre in 1998 and $134 per acre in 1999.

NewLeafY potato varieties are expected to provide benefits both in the production of certified seed potatoes as well as for commercial growers. Potatoes grown for seed are certified by state agencies against viral infection. The availability of NewLeafY potatoes eliminates the need for pest control targeted at the PVY, and allows growers to produce seed free of PVY.

Table : Insecticide Use in U.S. Fall Potato Production 1999

	Percent of Acres Treated	Total Applied (1,000 lbs)
Aldicarb	5	141
Azinphos-methyl	7	38
Carbaryl	3	30
Carbofuran	10	204
Cyfluthrin	3	**

Contd...

	Percent of Acres Treated	Total Applied (1,000 lbs)
Diazinon	2	31
Dimethoate	13	100
Disulfoton	1	19
Ethoprop	8	331
Endosulfan	16	221
Esfenvalerate	16	9
Fonofos	1	30
Imidacloprid	34	55
Malathion	*	2
Methamidophos	29	520
Methyl parathion	1	12
Oxamyl	2	13
Permethrin	8	15
Phorate	23	691
Phosmet	4	32
Piperonyl butoxide	2	22
Propargite	4	76
Spinosad	3	2
Total	93	2,596

*less than 1% of acres treated

** less than 1,000 lbs. applied

Source: USDA NASS

Table : Average Number of Insecticide Treatments for Conventional and NewLeafPlus Potatoes

	1998	1999
Russet Burbank	5.3	5.3
NewLeafPlus	1.0	3.2

Note: Results from field trials conducted in 8 locations in 1998 and 6 locations in 1999.

Source: Owens

SOYBEANS

Roundup Ready soybean varieties were introduced for planting by U.S. farmers in 1996, allowing treatment of a growing soybean crop with glyphosate (Roundup). Roundup Ready soybeans have been widely adopted by U.S. growers. Figure 10 shows adoption rates since 1996 in the U.S. By 2000, growers planted 54% of U.S. soybean acreage to glyphosate tolerant soybeans (USDA NASS 2000a).

Soybean is the second largest crop in the U.S. after corn, planted on 74.5 million acres in 2000 (USDA NASS 2000a). Acreage planted to soybean has expanded in recent years due to several factors. Improved yields through variety improvements, adoption of moisture-saving no-till practices, strong soybean prices relative to other crops, and elimination of acreage reduction programs are all factors that have contributed to expanded plantings. Total soybean crop value in 1999 was $13 billion (USDA ERS 1999).

Soybean acreage is centered in the Midwestern states, though 30 states have significant acreage planted to soybeans each year. Illinois and Iowa each plant over 10 million acres of soybeans. Other major soybean states include Minnesota, Indiana, Missouri and Ohio (USDA NASS 2000a).

The U.S. is the largest producer of soybeans in the world, growing nearly half of the total world soybean crop. Other major producing countries include Brazil, China and Argentina. The U.S. exports approximately one-third of its soybean production, primarily to Asia and Europe, which together account for over 70% of total exports. Competition in export markets comes from Brazil and Argentina, as China is a net importer of soybeans (USDA NASS 2000b).

The primary reason growers have adopted Roundup Ready weed control programs is the simplicity of a weed control program that relies on one herbicide to control a broad spectrum of weeds without crop injury or crop rotation restrictions. Before the introduction of Roundup Ready soybean varieties, growers would choose between many herbicides, often applying three or more active ingredients, some of which would cause damage to the growing soybean plants, or cause harm to corn crops that commonly follow soybeans (Gianessi, et al. 2000).

Roundup is a highly effective, broad spectrum herbicide that controls both broadleaf and grass weeds. Each year, state extension services release weed control guides for field crops including soybeans. The guides provide information on the efficacy of available herbicide treatments on specific weed species, as well as ratings of crop safety.

In the Michigan State University weed control guide, in which 182 treatments are rated on 24 different weed species, Roundup, used over Roundup Ready soybeans received 23 good or excellent ratings. In addition, the Roundup treatment is rated with a minimal risk of crop injury. The next best available treatment with similar crop safety received only 16 good or excellent ratings (Kells, et al.).

Growers also have more flexibility in timing herbicide treatments with the Roundup Ready system. Maximum weed heights at which Roundup is effective on most weed species are higher than other available herbicides. This allows growers to treat later if needed and still get effective weed control. Further, some commonly used soybean herbicides may cause injury to rotation crops. Because of this potential for injury to crops following soybeans, rotation restrictions are specified on the labels of these herbicides. For instance, sugarbeets may not be planted for 40 months after a field is treated with imazethapyr, a commonly used soybean herbicide.

Potential impacts of adopting Roundup Ready weed control programs include changes in costs, yields and pesticide use. Roundup Ready programs were introduced to be price competitive with existing conventional programs. The introduction of competitively priced Roundup Ready programs resulted in manufacturers of other products dropping their prices, in some cases by 40%. This resulted in an estimated $216 million cost savings for soybean growers in 1999 compared to 1995, the year before Roundup Ready varieties were introduced, including the technology fee paid by growers who planted Roundup Ready varieties.

Table : Soybean Weed Control Costs

millions

	1995	1998	1999
Herbicide Expenditures	$1,865	$1,482	$1,441
Technology Fee	$0	$160	$208
Net Weed Control Costs	$1,865	$1,642	$1,649

Note: Calculated assuming herbicide expenditures in 13 states represent 80% of U.S. total.

Herbicide use in soybeans has been affected dramatically by the introduction of Roundup Ready soybean varieties. The mix of herbicides being used in soybeans has changed. As one would expect, the use of glyphosate has increased, from being used on 20% of acreage in 1995 as a burndown or spot treatment, to being used on 62% of acres in 1999.

The use of other herbicides has decreased. Imazethapyr, the most widely used soybean herbicide in 1995, was used on 44% of soybean acres in 1995, compared to 16% in 1999. Table 29 shows use estimates of individual soybean herbicides in 1995 and 1999.

USDA pesticide use estimates show an increase in the total amount of herbicides used in soybean production, which is at least partially due to increased acreage. Herbicide application rates are the same in 1999 as in 1995.

Table shows applications rates for 8 states since 1990. Growers have reduced the number of herbicide applications. Comparing 1995, the year before Roundup Ready varieties were introduced, and 1999, the last year for which data are available, the number of herbicide applications decreased by 19 million, or 12%. These changes in herbicide use occurred even though the total number of soybean acres increased by 18% between 1995 and 1999. The decrease in herbicide applications demonstrates growers using fewer active ingredients and making fewer trips over the field, which translates into ease of management.

The benefits of the introduction of Roundup Ready soybeans include a cost savings of $216 million in annual weed control costs and 19 million fewer soybean herbicide applications per year.

Table : Soybean Herbicide Use 1995 and 1999 Active Ingredient

percent of acres treated

	1995	1999
2,4-D	10	5
2,4-D,B	1	*
Acifluorfen	12	3
Alachlor	4	2
Bentazon	12	4
Chlorimuron	16	12
Clethodim	5	5
Clomazone	4	1
Cloransulam	5	
Dimethenamid	1	*
Ethalfluralin	1	*
Fenoxaprop	6	4
Fluazifop	10	4
Flumetsulam	2	2
Fomesafen	4	4
Glyphosate	20	62
Imazamox	3	
Imazaquin	15	5
Imazethapyr	44	16
Lactofen	5	2
Linuron	2	*
Metolachlor	7	4
Metribuzin	11	5
Paraquat	2	1
Pendimethlin	26	14
Quizalofop	6	1
Sethoxydim	7	3
Sulfentrazone	4	
Thifensulfuron	12	5
Trifluralin	20	14

Source: USDA NASS

* less than 1%

Table : Soybean Herbicide Application Rates (AR, IA, IL, IN, MN, MO, NE, OH)

Year	Rate (lbs/acre)
1990	1.26
1991	1.23
1992	1.11
1993	1.06
1994	1.11
1995	1.00
1996	1.05
1997	1.18
1998	1.06
1999	1.00

Source: USDA NASS

SUMMARY

The reasons U.S. growers are adopting genetically modified crops vary. The fact that new genetically modified varieties are capturing a large portion of the very competitive pesticide markets for major crops indicates that the new technology is delivering benefits to growers.

Bt corn varieties have provided control of a destructive insect pest that was mostly uncontrolled previously, which has resulted in increased yields. Cotton growers have adopted varieties that are insect and/or herbicide tolerant, which has allowed reduced pesticide use, increased yields and increased grower returns. Potato growers have not yet widely adopted Bt varieties. Soybean growers have adopted Roundup Ready soybeans largely due to the simplicity and efficacy of weed control provided by the program. The result has been soybean herbicide cost savings and a significant reduction in the number of soybean herbicide treatments.

11

Agricultural Biotechnology: Benefits of Transgenic Soybeans

INTRODUCTION

Soybeans and other crops have been improved genetically for many decades through traditional crop breeding – a technique that requires that species be sexually compatible. With the development of biotechnology methods, scientists have the ability to transfer single genes from one living organism into another, regardless of species or sexual compatibility. Varieties that are developed through the transfer of genes between species that are not sexually compatible are referred to as "transgenic." Transgenic soybean plants have been developed with a gene from a soil bacteria that allows the use of an herbicide that would normally kill soybeans. Transgenic soybean varieties have been adopted rapidly by U.S. growers (Table 1). Transgenic soybean acreage represents approximately 60 % of the total acreage planted to transgenic crops grown in the U.S. (Corn and cotton are the other major crops with transgenic variety plantings in the U.S.)

Farmers are planting transgenic soybeans in the U.S. because of the weed control benefits that result from the use of an herbicide that damages conventional varieties. This herbicide, glyphosate (trade name "Roundup") is a cost effective solution to many of the weed problems that U.S. soybean growers must overcome every year. The herbicide tolerant transgenic soybean is referred to as "Roundup Ready." The purpose of this report is to describe and quantify the weed control benefits provided on soybean acreage planted with the transgenic varieties in 1998.

Table : Herbicide Tolerant Transgenic Soybean Acreage: U.S.

Year	# of Acres (Millions)	% of U.S. Acreage
1996	1	2
1997	9	13
1998	27	37
1999	35	47

Source: Monsanto

U.S. SOYBEAN PRODUCTION

Soybean is one of the three largest crops grown in the U.S., grown on approximately 70 million acres in 1999 an area smaller than corn and comparable to wheat. Thirty states have significant soybean acreage. Soybean production is centered in the Midwest where ten states account for 73% of U.S. acreage and production. The states of Illinois and Iowa each account for more than 10 million acres of soybeans. The Delta states of Mississippi, Arkansas and Louisiana together account for 10% of U.S. acreage of soybeans.

Soybeans were grown primarily as forage crops in the U.S. through the 1930's. The production of soybeans increased in 1934 in response to the severe drought of that year in the upper Midwest. The performance of soybeans under drought conditions was better than corn, thereby enticing farmers to try this relatively new crop. Soybean harvested for seed represented 40% of the planted acreage in 1939, indicating expanding acreage for processing.

Prior to World War II, the U.S. imported 40% of its edible fats and oils. At the advent of the War, this supply was cut, and processors turned to domestically produced soybean oil. By 1944, 72% of the planted soybean acreage was harvested for seed [100]. World demand for cooking oil, salad oil and red meat increased substantially immediately after World War II.

These demands stimulated the rapid expansion of soybean production in the U.S.. In the 1950's, soybean meal became available as a low-cost, high-protein feed ingredient, triggering explosive growth in U.S. livestock and poultry production. U.S. soybean acreage increased from less than 20 million acres in the early 1950's to over 50 million acres in the 1970's and a record 72 million acres in 1998, representing 27% of the acreage planted to all crops in the U.S.

Total annual U.S. soybean production increased from about 500 million bushels in the 1950's to over 2.5 billion bushels (150 billion pounds) in 1997. Figure 2 charts U.S. soybean production 1944-1999. The increase in annual total U.S. production resulted not only from the expansion of acreage but also from a steady rise in average U.S. soybean yield per acre.

Two-thirds of the increase in soybean yields in the U.S. from 1943 to 1960 is attributed to the planting of improved varieties, with increased fertilizer use accounting for the remaining one-third of the increase.

A statistical analysis of the change in soybean yields from 1965 to 1979 concluded that weed control provided by the use of herbicides accounted for 62 percent of the yield increase while further variety improvements accounted for 13 percent of the increase.

In the 1990's, U.S. soybean farmers received an average of $6 per bushel. However, prices dropped to $5/bushel in 1998. The total value of the soybean crop was $14 billion in 1998, representing approximately 15% of the value of all crops grown in the U.S. The U.S. produces nearly half of the total world soybean crop. Other major producing countries include Brazil, China and Argentina. Competition in export markets comes from Brazil and Argentina as China is a net importer of soybeans. The U.S. accounted for 60% of world exports of soybeans in 1997/98.

About 28% of U.S. soybeans (valued at $4.8 billion) were exported in 1998. The U.S. exported soybean meal and soybean oil valued at $1.6 billion and $0.9 billion, respectively, in 1998.

SOYBEAN PRODUCTS

Approximately 74 billion pounds of soybean meal and 18 billion pounds of soybean oil were produced by U.S. processors in 1998. (See Appendix 1 for a description of soybean processing.) Ninety-six percent of the soy meal produced in the U.S. is used as an animal feed. Three percent of the soybean meal is used in human food, and approximately 1% is used in industrial uses.

Soybean oil makes up about 75% of the oil used in U.S. shortening (baking and frying fats), margarine and salad/cooking oil. The large volume of soybean oil use has been attributed to at least three factors: 1) a plentiful and dependable supply; 2) competitive price; and 3) the improvements made in the flavor and oxidative stability of the oil.

Until 1980, cottonseed oil was the preferred oil for mayonnaise products. However, the cottonseed oil emulsion in mayonnaise is rather weak and sometimes breaks down at refrigerator temperatures. With improved flavor stability and stronger emulsion of soy oil, all mayonnaises produced commercially in the U.S. are now prepared exclusively with soybean oil as the key component. Similarly, soybean oil is now used almost exclusively in all prepared salad dressings and imitation mayonnaises sold in the U.S..

Lecithin is a co-product of degumming soybean oil, that, after processing, has significant commercial value as an emulsifier in foods. Soy lecithin is normally added to such food products as shortening, margarines, baked goods, chocolate confectionery coatings, peanut butter, powdered mixes and dietary food. Lecithin increases the stability of these products.

When added to margarine, lecithin prevents "bleeding" of the moisture present and reduces spattering during frying. In baked goods, lecithin functions as a useful emulsifier. It helps bring about rapid and intimate mixing of the shortening in the dough. In making chocolate, about 0.25-0.35% lecithin is added. It reduces viscosity of chocolate markedly, enables the manufacturer to apply a uniform coating, decreases the time for grinding and mixing the various ingredients and produces a more stable chocolate. Lecithin is used to keep the chocolate and cocoa butter in a candy bar from separating. In peanut butter, added lecithin, normally at 1-2% levels, gives a smoother, creamier spread that does not separate during wide temperature variations.

Approximately 45% of the soybean meal used as feed in the U.S. is for poultry feed while swine and beef feed represent 27% and 9%, respectively. Pet food, dairy animals, egg hatcheries and aquaculture make up the remainder. A typical broiler feed may contain 17-22% protein, of which more than 90% usually comes from soybean meal. The most important factor responsible for increasing the feed efficiency in poultry production has been the utilization of a high-protein diet containing dehulled and defatted soybean meal. The efficiency in broiler production increased from a gain of about 4 kg of feed per kg of bird in 1948 to the present rate of 2 kg of feed per kg of bird.

Soy protein products include defatted soy flakes, soy meal, soy flour and grits, soy concentrates, soy isolates, texturized soy proteins, full-fat soy flour and enzyme active soy flour. Soy flour or grits are used as an ingredient in a variety of food products, including soups, stews, beverages, desserts, bakery goods, breakfast cereals and meat products. Soy flour and concentrates may be processed further by thermoplastic extrusion to impart meat-like texture to these products. Soy protein ingredients are used extensively in meat products as extenders.

In bakery products, soy proteins are used as replacement for more expensive egg whites. Soybean proteins are used in whipped toppings, ice creams, coffee whiteners, imitation milk and cheese, frozen desserts and yogurts. Soy isolates are used in most of these products because of their high protein content and ability to impart desirable functionalities. Soy flours, concentrates and isolates are used as a source of protein in various infant foods based on cereals, vegetables and meats. Soy flour use in some products limits fat absorption during deep frying, for example, in doughnuts. Soy hulls are processed into fiber bran breads, cereals and snacks.

Table : Fats and Oils Used in Edible Products in the U.S.: 1997/1998

(Million LBS/YR)

	% Soybean Product	Total Soybean Oil	Of Total
Shortening			
(Baking and Frying Fats)	5,183	4,247	82
Margarines	1,549	1,474	95
Salad and Cooking Oil	6,803	5,673	83
Other Edible	306	65	21

Source: October 1997 – September 1998.

SOYBEAN PHYSIOLOGY

Soybean is a valuable agricultural commodity due to its unique chemical composition. The soybean seed consists of lipids, proteins, carbohydrates and minerals. Other valuable components found in soybeans include phospholipids, vitamins and minerals. On average, oil and protein together consitute about 60% of dry soybeans. The remaining dry matter is componsed mainly of carbohydrates (35%) and ash (5%). Nine fatty acids make up the lipid content of soybeans [98]. During seed development, soybeans store their lipids, mainly in the form of triglycerides, in organelles known as oil bodies. The soybean contains 20% oil, the second highest oil content among food legumes. The highest oil content is found in peanut, with about 48% oil; the third highest oil content is found in chickpeas, with about 5%. Crude soy oil contains impurities such as phosphatides, free fatty acids, gummy substances, color bodies, hydrocarbons and a waxy fraction. Refining operations remove these impurities. Nineteen amino acids make up the protein content of soybean flour, including the essential amino acids, lycine, isoleucine, leucine and valine [98]. Among cereal and other legume species, soybean has the highest protein content, about 40%. Other legumes have a protein content between 20 and 30%, whereas cereals have a protein content in the range of 8-15%.

Some of the proteins in soybeans are enzymes. The enzymes in soybeans have catalytic activity. Catalysts increase the rate at which chemical reactions take place, but they do not cause a reaction to take place. They control the rate at which chemical reactions take place in the plant. Each enzyme is specialized in the reaction that it catalyzes [94]. For example, certain enzymes act as catalysts in the soybean plant's production of the essential amino acids.

A typical soybean plant cell contains at least 10,000 different enzymes. As long as conditions are favorable, enzymes will catalyze repeatedly their respective chemical reactions.

SOYBEANS – AGRONOMIC FACTORS

The bushy green soybean plant is a legume related to clover, peas and alfalfa. Each soybean plant produces 60 to 80 pods, each holding three to five pea-size beans. Soybeans are generally planted in May and June. The soybean plant flowers and pod filling occurs in July and August. Harvest begins in September and is largely completed by mid-November.

Only 10% of U.S. soybean acreage is typically planted to continuous soybeans. Fifty-eight percent of the nation's soybean acreage is in an annual rotation with corn while 20% is rotated annually with other row crops or small grains. Five percent of the soybean acreage is idle or fallow during the previous year. Seven percent of the nation's soybean acres is double cropped annually with soybeans being planted after winter wheat is harvested.

Soybeans are planted at a high per acre plant density with 84% of U.S. acreage planted with 75,000 or more plants per acre. The average plant population on U.S. soybean acreage is 145,000. The average row spacing in U.S. soybean fields was 20 inches in 1994. Thirty-six percent of U.S. soybean acreage is planted in rows spaced less than 10 inches apart.

Conservation tillage programs, which reduce erosion from growers fields, are widely used by soybean growers. In 1994, 50% of soybean acreage was under conventional tillage and 50% was in conservation tillage programs. Conservation tillage programs where at least 30% of the soil surface covered with crop residue, include mulch till, ridge till and no-till systems. Twenty-four percent of the acreage was in mulch till, 1% was in ridge till and 25% was in no-till in 1994..

Typically, less than 1% of U.S. soybean acreage is treated with insecticides or fungicides. Soybean plants have been bred with considerable insect and disease resistance. Another factor that reduces the impact of insect feeding is the ability of the soybean plant to compensate for a significant amount of defoliation. The typical soybean farm operator in Central Illinois in 1998 grew 449 acres of soybeans and 455 acres of corn.

SOYBEAN GENETIC IMPROVEMENTS

The soybean is a native of eastern Asia, where it is known to have been cultivated for over 4,000 years. Soybeans were first grown in the U.S. in about 1804 in the state of Pennsylvania, but they were still a very minor crop 100 years later. No more than eight soybean cultivars were grown in the U.S. prior to 1898. From 1898 to 1923, more than 1,000 cultivars were introduced – most sent by research stations or grain merchants in Asia or brought in by agricultural explorers, diplomats or other travelers. As a result of the increasing success of soybeans, the USDA sent plant explorers to Asia, and from 1924 to 1933, 6,651 soybean accessions were introduced to the U.S.. By 1947, over 10,000 soybean varieties had been introduced by the USDA.

Genetic improvement in soybean has resulted from the continual development and release of new cultivars with greater genetic yield potential than their predecessors. The amount of genetic improvement in yield that has been realized by hybridization and selection has been substantial. For example, a 25 % increase in yield occurred between plant cultivars released before 1940 and cultivars released after 1970. Other characteristics that have been the focus of traditional crop breeding include lodging resistance, reduced plant height, seed size, seed quality, oil quality, shattering resistance and resistance to insects, diseases and nematodes.

All soybean cultivars now grown in the U.S. were developed by artificial cross pollination.

Soybeans have a very limited tolerance to many of the herbicides that have been used to kill weeds in the crop. Rates of many herbicides are kept low because of the limited ability of the soybean plant to detoxify the chemicals if used at higher rates. Soybean plants are often damaged by herbicides under varying environmental conditions, such as higher than normal rainfall.

Soybean breeders traditionally have placed a relatively low priority on enhancing herbicide resistance.

In order for traditional breeding to produce herbicide resistant soybeans, genes for resistance must be available in crossing material that is compatible with soybeans. A lack of sufficient variability in resistance levels in soybeans has hindered breeding efforts. Soybean cultivars frequently show differences in the degree of injury caused by herbicides, but the occurrence of genotypes giving highly sensitive herbicide resistance has been rare. This requirement has been the limiting factor for producing herbicide resistant soybeans with naturally occurring genes.

However, there has been limited work in genetically improving soybean cultivars for herbicide tolerance. Soybean cultivars have been "selected" for increased tolerance to the herbicide metribuzin. This selection procedure did not entail the creation of a new soybean cultivar; merely the selection among the populations of existing cultivars of a more tolerant plant.

Two successful attempts to create genetically altered soybean plants with herbicide tolerance have been made: (1) the use of mutation breeding to develop soybean cultivars with increased tolerance to sulfonylurea herbicides and (2) the creation of transgenic soybean cultivars with tolerance for use of the herbicide glyphosate. These two genetic improvements are described further in this section.

Once the genetic change was accomplished in a single soybean cultivar, traditional crop breeding methods were used in order to increase the number of soybean cultivars with the tolerance to sulfonylureas and to glyphosate. Because of the importance of cross breeding in improving soybean cultivars, this section includes a discussion of these traditional methods. Since an understanding of crop breeding is dependent upon understanding the reproductive process of soybean plants, the next section of this report discusses that phenomenon.

Reproductive Process

The soybean flower is a perfect flower, meaning that both the male and female parts are present in the same flower. When the pollen is shed, it drops immediately on the female part of the same flower. This often happens before the flower opens (blooms). The pollen produced within the flower fertilizes the ovary of the same flower. Because of the structure of the floral parts, soybeans are almost completely self-fertilized. Crossing has been estimated to be less than 0.5 %. Soybean pollen is too heavy for wind transport but can be transported by insects. Enclosed within the petals are ten anthers (male) arranged in a circle around a pistil (female). The pistil has a stigma at its tip that receives the male sex cells (pollen) from the anthers. Pollen germinates on the stigma, developing tubes that then grow into the stigma. Only a few pollen tubes reach the locule and compete for ovules to fertilize. Finally, the pollen tube grows into the ovule. Here the pollen tube tip bursts and releases two sperm cells. One sperm fuses with the egg and forms the first cell of the embryo while the other sperm fuses with the secondary nucleus forming the primary endosperm nucleus.

The time from pollination to fertilization varies from 8 to 10 hours. From the moment of fertilization, the ovary starts developing into the fruit.

Artificial Cross Breeding

The soybean cultivars grown by farmers in the U.S. up to the 1940's were introduced from Asia. Introductions were tested for agronomic performance, and the superior ones were released to farmers.

The best introductions were used as parents to develop superior cultivars, which are ancestors of current cultivars. In the mid-1930's, soybean breeders began making two-way crosses between the best yielding introductions from Asia and produced progeny that were superior in yield to either parent. Subsequent matings among these selections generated a new round of superior cultivars released in the 1950's and 1960's.

The stigma is receptive to pollination at least one day before the anthers are sufficiently developed to shed pollen. This lag time between female and male development permits the breeder to introduce pollen from an outside source to obtain hybrid seed. To prepare the flower for pollination, the five lobes of the calyx and the petals are removed to expose the female and male organs. Emasculation is not necessary. Apparently an exposed stigma that is not promptly pollinated dries up and becomes non-receptive before anthers from the flower can shed pollen.

Flowers of the female parent are pollinated artificially the day before their male parts have matured enough to shed pollen. A floral bud at the appropriate stage is swollen and the corolla is visible. The five sepals of the calyx are removed with tweezers to expose the corolla. The male parts of a matured donor plant are removed by use of a pair of tweezers. The pollen is placed on a stigma by brushing the anthers against it. If the pollination is successful, a pod will be visible in about seven days. Breeding soybeans is tedious because the flower is very small. The soybean flower is one-quarter inch long.

The number of crosses a person can make in a day ranges from less than 50, by inexperienced personnel, to about 150, by individuals with considerable experience. One half or more of the pollinated flowers may abort.

Cross breeding is initiated by making a cross between two parents to produce a hybrid. The hybrid plant is crossed back to the parent that is being improved, called the recurrent parent because it recurs, or is used repeatedly for crossing. The other parent is the donor parent and contributes a desirable gene; however, the donor parent is not used in the backcrossing program. The purpose of crossing back to the recurrent parent is to recover all of its desirable genes. Each time a cross is made back to the recurrent parent, an additional 50 percent of its genes are recovered. The breeder will continue backcrossing until the desirable level of genes from the recurrent parent has been recovered. Although more than 10,000 soybean strains have been introduced into the U.S. since the early 1900's, only a limited number of these introductions form the genetic base for cultivar development for the hybridization programs.

An analysis of the pedigrees of 136 soybean cultivars that originated from hybridization breeding programs during 1939 to 1981 revealed that only a limited number of ancestral introductions contributed germplasm. Only five introductions were the cytoplasm source for 121 of the 136 cultivars. The ancestry of the nuclear material in these 136 cultivars was narrow, with 12 introductions contributing about 88 percent of the germplasm. Traditional breeding procedures that emphasize the mating of elite strains are, in effect, continually recombining the genes contributed by a limited group of ancestral introductions.

The reason for such a narrow genetic base is that certain crosses of many of these ancestors were particularly productive in generating high-yielding cultivar releases, that, in turn, served as the parents in subsequent rounds of mating, resulting in still higher yielding releases.

Exotic soybean germplasm has not been used to a great extent in commercial breeding programs primarily because soybean breeders have been successful in using crosses of elite strains to generate high-yielding cultivars, and because crosses involving at least one unadapted parent have a low frequency of superior segregates in their progeny.

Mutation Breeding

Mutations are sudden changes in the hereditary material of an organism. Regarded broadly, they include all such changes that cannot be accounted for by the normal recombination of the units of heredity.

Mutations can result from the loss or rearrangement of a part of a chromosome or from the loss of entire chromosomes. Mutations also include the duplication of chromosome parts and the occurrence of additional chromosomes. Parts of chromosomes may change places. A piece of broken chromosome may rotate 180 degrees before rejoining the chromosome. Perhaps a broken piece is lost altogether along with a consequent loss of the genetic information that it carried.

Natural mutations have a number of causes, including cosmic rays, heat and aging, although in most cases the origin of a specific spontaneous mutation is unknown. Mutations occur spontaneously in all living things. By changing the chemistry of the organism's genetic material or altering the structure of a chromosome, a mutation changes the structure or function of the organism and its offspring. Most often the mutations are harmful. Once in a while a mutation is beneficial, increasing the organism's chance to survive and reproduce or, in the case of plants that are directly useful to humans, increasing its value.

On this foundation, a number of plant breeders have ventured to induce mutations artificially and, thus, to take advantage of the beneficial ones. The principal mechanisms for inducing mutations is the irradiation of seeds, although mutations are also produced by irradiating entire plants and by treating seeds with mutagenic chemicals.

More than 1,000 varieties of induced-mutant crop plants – including wheat, rice, barley, oats, soybeans, fruit trees and ornamental plants – have been released to growers and are being grown by farmers on millions of acres throughout the world. A remarkable example has been the mutation breeding work of chrysanthemum in The Netherlands. Within a few years after an irradiation project had started, a whole range of floral color materials were obtained that quickly replaced other cultivars. Swedish scientists were pioneers in the practical uses of induced mutations and developed a great deal of the present technology with mutation breeding of barley.

When breeders induce mutagenesis, they are likely to discover useful mutations that do not exist in naturally evolved populations. However, the induction of a specific desirable mutation remains a random process. Even with the help of mutagens, any particular mutation occurs rarely.

A mutagen does not produce a specific mutation. What is achieved is an increase of the general mutation frequency, and, thereby, increased likelihood of an agronomically valuable mutation.

Ionizing radiations are available to plant breeders from a variety of sources: x-rays, gamma rays from radioactive isotopes (chiefly cobalt 60 and cesium 137) and neutrons from nuclear reactors.

Chemical mutagens are more readily available to plant breeders than radiation sources, and the ratio of mutational to undesirable modifications is somewhat better for chemical mutagens than for irradiation. Hence, chemical mutagens are becoming increasingly popular. For practical plant breeding, the most interesting group of chemical mutagens are alkylating agents. These compounds bear one or more reactive alkyl groups capable of being transferred to other molecules at positions where the electron density is sufficiently high. The alkylating agents react with DNA by alkylating the phosphate groups as well as the purine and pyrimidine bases. Alkylating agents produce unspecific effects, i.e., both small effects like point mutations and major chromosonal aberrations are induced. Among the 30 to 40 chemical mutagens in the category, one of the most powerful and useful mutagens is ethyl methanesulfonate (EMS). The mutation rate with the use of EMS is approximately 50%.

A soybean seed contains a mature, arrested embryo. The embryo in a mature seed of soybeans has already initiated its first leaves. Mature seeds are metabolically quiescent. As long as the seed is relatively dehydrated, it remains dormant. It is alive, but all metabolic processes are occurring at very slow rates. Most of the common mutagens work by encouraging mistakes during DNA replication. Effective mutagenisis depends on applying the mutagenic agent at exactly the time when DNA is replicating. To activate the plant embryo, seeds need only to be saturated with water.

In the 1980's, the Dupont Company conducted a crop breeding program to enhance soybean tolerance for sulfonylurea herbicides. The molecular target of the sulfonylurea herbicides is the enzyme ALS, also known as acetohydroxyacid synthase (AHAS), that catalyzes the first common steps in the biosynthesis of the branched-chain amino acids isoleucine, valine and leucine in plants and microbes. This pathway is not present in animals [96].

Seed mutagenesis (using N-nitroso-N-methylurea and ethyl methanesulfonate) followed by selection for resistance to sulfonylurea herbicides yielded a soybean mutant with a high degree of resistance to applications of a variety of sulfonylurea herbicides. Four hundred fifty thousand soybean seeds were presoaked in tapwater. The swollen seeds were then soaked in the mutagen. Treated seeds were washed and field planted. Seeds of these plants were harvested and were screened for resistance to sulfonylureas by soaking them in a solution that included a sulfonylurea herbicide. These seeds were then planted and 75 mutants were selected based on their ability to form true leaves within several weeks. (This treatment severely retarded development of the parent soybean cultivars.) One mutant was significantly more tolerant of sulfonylurea treatments.

The mutants have an altered ALS that shows reduced sensitivity to inhibition by sulfonylureas. Tolerance was conferred by a mutation in a single gene. The sulfonylurea tolerant soybeans are designated by the trademark "STS." In the late 1980's, Dupont released the "STS" germplasm to private soybean seed companies to breed into their soybean lines. STS soybeans were introduced commercially in 1992. Dupont markets "Synchrony STS" for use with STS soybeans only. Synchrony STS consists of chlorimuron and thifensulfuron.

Sulfonylurea-tolerant soybean cultivars are more tolerant of higher rates of the sulfonylurea herbicides chlorimuron and thifensulfuron. These higher rates are necessary for controlling difficult weeds, including lambsquarters, pigweeds, morningglories, sicklepod and velvetleaf.

Experiments in 1993 compared tolerance of a non-sulfonylurea tolerant soybean with an STS variety when treated with sulfonylurea herbicides. Injury to the non-STS variety ranged from 13 to 65% and averaged 3% to the STS variety with the same treatments.

Transgenic Plants

Advances in biotechnology have made it possible to transfer genes from organism to organism by means that bypass the normal sexual processes governing intraspecific inheritance. Isolated genes are moved into a crop plant in such a way that the genes are integrated into the chromosomes and expressed. A significant advantage is the fact that whole plants expressing a foreign gene can be regenerated from single transformed cells. After being absorbed by plants, the herbicide glyphosate inhibits the enzyme 5-enolpyruvylshikimate-3-phosphate synthase (EPSPS). Glyphosate binds to EPSPS resulting in EPSPS's inhibition, causing the plant to starve for EPSP and the metabolic products derived from EPSP.

Research to develop glyphosate tolerant plants began at Monsanto in the early 1980's. Simple selection as whole plants or cell cultures either with or without mutagenesis agents proved to be largely unsuccessful. Attempts to identify glyphosate tolerant plant EPSPS as well as extensive random mutagenesis efforts failed to identify EPSPS with adequate glyphosate tolerance.

The ability to metabolize glyphosate is distributed widely among soil bacteria. Research demonstrated that EPSPS from a number of bacteria exhibited tolerance to glyphosate. Monsanto scientists collected bacterial cultures from diverse sources and analyzed them for their tolerance to glyphosate. One type of bacteria represented heavily in the screening was glyphosate degrading bacteria. The rationale being that perhaps organisms that can grow in the presence of glyphosate and degrade the herbicide may also express naturally glyphosate-tolerant EPSPS. The EPSPS with the highest tolerance to glyphosate found in the screening was CP4 EPSPS from agrobacterium tumifacien, that demonstrated extremely high glyphosate tolerance.

The gene from CP4 EPSPS was cloned and introduced into soybeans by the particle acceleration method. Soybean cultivar A5403, a commercial variety developed by ASGROW Seed Company was used for the transformation.

The particle gun can accelerate microscopic projectiles to initial velocities of about 1,400 feet per second. At that speed the projectiles penetrate the cell walls of intact plant cells and carry DNA into the cell. With each bombardment, thousands of particles are accelerated at the same time, thereby delivering DNA into many cells simultaneously. Gold beads approximately 1.3-3 microns in diameter were used. Gold is a dense chemically inert material, that is non-detrimental to the plant tissue. Particle acceleration is capable of delivering biologically active foreign DNA into viable plant tissues at a frequency high enough to allow the recovery of stable transformations.

Immature soybean embryos were the targets of the rapidly accelerated DNA coated gold particles. Five to seven immature embryos were arranged on a plate and were subjected to particle acceleration. Effective penetration was limited to a depth of five to seven cell layers. Treated embryos were chopped and plasmolized, and protoplasts were isolated. The embryonic axes were then plated on a germination and shoot multiplication medium and transformed plants were recovered.

The lead soybean line with a Roundup Ready gene is denoted 40-3-2, and expresses the CP4 EPSPS gene product. Upon glyphosate treatment the transgenic plant remains unaffected because the continued action of the introduced glyphosate-tolerant EPSPS enzyme meets the plant's need for aromatic amino acids. The endogenous EPSP is inhibited by glyphosate (upon glyphosate treatment), however, the plant relies on the introduced glyphosate-tolerant EPSPS for EPSP synthesis.

Line 40-3-2 and progeny from crosses between line 40-3-2 and other soybean varieties were yield tested under weed-free conditions in 1992 and 1993. Data from these experiments showed that there is no yield penalty observed upon glyphosate treatment of this line with Roundup even at rates as high as twice the level to control most weeds.

Line 40-3-2 has been used in various breeding programs to develop new cultivars with a Roundup Ready gene. (Roundup Ready is Monsanto's trademark for its genes conferring glyphosate tolerance.) As a single dominant gene, the glyphosate-tolerant gene can be used very effectively in breeding programs.

Over 150 seed companies offer more than 1,000 Roundup-Ready varieties.

WEED COMPETITION IN SOYBEANS

Weeds compete with soybeans for soil moisture, nutrients, sunlight and space in the field. One cocklebur may occupy four to eight square feet of area, thereby reducing the space available for soybean growth. When weeds shade the soybean plant, less sunlight is available for soybean production. Most of the soybean yield reduction from velvetleaf and pigweeds is ascribed to shading by the weed leaves above the soybean canopy. As a result of competing with soybeans, uncontrolled weeds decrease the quantity of soybean seeds produced.

The efficiency of operation of harvesting equipment is also reduced by the presence of significant numbers of weeds. The quality of the harvested crop is directly impacted by weeds. Increases in moisture content, foreign matter, and splits have been documented when high levels of weeds are present at harvest.

Two of the factors that contribute to the strong competitive nature of weeds include high seed production leading to high plant density and a long duration in the soil. Weed seeds remain viable (capable of germination) for varying periods of time. Seed longevity represents a major survival mechanism for weed species; it constitutes a continuing source of emerging weeds in croplands.

As can be seen, the seeds of these species can survive in the soil for eight to forty years. Weed species re-infest the soil primarily due to the large amounts of seeds produced by a single plant. A single cocklebur plant can produce 900 seeds while a single pigweed plant produces 117,000 seeds.

In Minnesota, weed seed counts at four different locations in twenty-four different plots showed from 98 to 3,068 viable weed seeds per square foot of soil six inches deep – that converts to between 4 million and 133 million seeds per acre. In western Nebraska average cropland soil contained 200 million seeds per acre. The number of weed seeds that germinate and emerge in any given year is quite low in relation to the total number of seeds present – perhaps only 5 – 10% of the total seed population. A very high percentage of the total weed seed population in the soil survives from one year to the next.

Numerous reports in the literature quantify the effects of full- or partial-season weed interference on soybean seed yields. The data show considerable differences among species in interfering ability. Common cocklebur is the most interfering weed: 1 plant/m^2 depresses soybean yield by 30% while nine plants/m^2 depresses soybean yield by 80%. A less competitive weed species for soybeans is giant foxtail: 16 plants/m^2 depresses soybean yield by only 10%.

There are three categories of weeds: dicots, monocots and sedges. Dicotyledons (dicots) contain two cotyledons (seed leaves) while monocotyledons (monocots) contain only one. These two large groups of plants typically have different types, arrangements and locations of organs. Dicots have broad leaves with veins radiating from a midvein, a taproot and/or fibrous root system and flower parts in multiples of four or five. In contrast, monocots generally have long, narrow leaves with parallel veins, a fibrous root system and flower parts in multiples of three. Because of pronounced structural differences between dicots (broadleaves) and monocots (grasses), weed control methods can often be targeted specifically at one of these groups.

More than thirty plant species infest soybean fields in the major soybean producing areas of the U.S.. Annual broadleaf and grass weeds are major problems. In some areas, perennial grass, broadleaf weeds and sedges are troublesome. Table 4 lists important weed species in soybean fields in two states, representative of the Midwest (Illinois) and the Delta (Arkansas). As can be seen, a combination of broadleaf and grass weed species infest a sizable portion of the soybean acreages in both states. Ragweed, foxtail, nightshade, lambsquarters, smartweeds, and velvetleaf infest more acreage in the Midwest than in the Delta. Morningglory, barnyardgrass, signalgrass, prickly sida, hemp sesbania and sicklepod infest more acreage in the Delta. Cocklebur, johnsongrass, crabgrass and pigweeds are estimated to infest sizable acreages of soybeans in both soybean producing regions.

Two to four species typically dominate the soil seed population. In a typical field in the Midwest, weed control strategies are generally planned based on two grass weed species and three to five broadleaf species. Natural weed populations in most fields are high enough to cause devastating soybean yield losses if left uncontrolled. Loss figures of 50 – 90% are common for soybeans grown in natural weed infestations.

Research has shown that a period of 4 to 6 weeks without weed competition at the beginning of the growing season will allow production of maximum yields under most environmental conditions. Any weed emerging in the crop after this initial weed free period will not compete effectively with soybeans and will not affect yield potential due to the soybean canopy which shades emerging weeds. Similarly, a period of 4 to 6 weeks of weed interference at the beginning of the season usually can be tolerated by soybeans with no significant yield loss provided that the crop is maintained weed free for the remainder of the season.

WEED CONTROL IN SOYBEANS: 1940'S-1950'S

In the 1940's and 1950's, tillage was the primary method used to control weeds in U.S. soybeans. The use of several shallow cultivations prior to planting soybeans to destroy as many annual weeds as possible was regularly recommended. Various implements, such as the rotary hoe and spike-toothed harrow, were used after planting to perform shallow tillage to uproot very young annual weeds between the soybean rows. Weed control by the tillage method is achieved primarily by (1) the burial of small annual weeds in soil thrown over them through the action of tillage tools and (2) the disruption of the intimate relationship between the weed plant and the soil, whereby (a) the soil is loosened about the roots, resulting in disruption of water absorption and death by desiccation, or (b) the plant is "cut off" below ground. Care must be taken in the tillage operation so as not to injure the roots or aboveground parts of the crop plants.

Research demonstrated that rotary hoeing provided 70-80% weed control three to five days after soybean emergence with two repeat treatments at about five-day intervals. However, untimely rotary hoeing applied when weed seedlings were bigger (in the 1-3 leaf stage) decreased weed control effectiveness to 50%. The rotary hoe should travel at a relatively high speed (8-12 mph). The rotary hoe is most effective on small weeds. For bigger weeds, growers relied on the use of slower-moving shovel cultivators.

Researchers in the 1950's reported effective control of weeds between the rows with sweep cultivation and some control of weeds in the row by throwing soil into the row during cultivation. Throwing soil into the row is risky since it may cause excessive ridging of soils in the row.

Timely weed removal treatments with the rotary hoe were sometimes difficult to apply because of wet conditions – mud would ball up excessively on the rotary hoe. Consequently, rotary hoeing was not always practiced when weeds were at an optimum size. To reach a maximum effectiveness, the hoeing must be done before the weeds reach a height of more than a quarter inch. Prolonged rainy periods often delayed the use of the rotary hoe in farmers' fields beyond the time when it is effective. If rotary hoeing is delayed, the weeds develop extensive root systems preventing their removal with the implement.

The best weed removal system for soybeans in the 1950's was determined to be two timely rotary hoeings plus two shovel cultivations. Even with timely use of cultivation, soybean yields were reduced because the tillage operations did not control effectively the weeds growing in the row with the soybeans. An eleven year experiment (1952 – 1962) in Iowa estimated soybean yield reductions resulting from weed infestations which survived good cultural and mechanical weed control methods. Average soybean yield reduction was 10% despite best mechanical weed control practices. Uncontrolled weeds were confined to a four to six inch band centered on the soybean row.

In the 1950's soybean farmers in the Midwest generally cultivated two to three times. In 1964, soybean acreage in Illinois was cultivated two times with a shovel or sweep cultivator and an additional one time with a rotary hoe or harrow cultivator. USDA estimated that the average annual national loss in the potential production of soybeans due to weeds was 17% for the period between 1951 and 1960. The USDA loss estimate includes a yield loss of 14% and a loss of three percent in quality due to weed seed dockage, damage in cleaning to remove weed seeds, split beans due to presence of weed seeds and off flavors. The need to cultivate soybeans for weed control was seen as the limiting factor as far as the number of acres that one person could manage since planting had to be stopped so that cultivating could begin.

Each tillage operation reduces the amount of crop residue and clods on the soil surface and this, in turn, increases soil erosion susceptibility.

HERBICIDES – AN OVERVIEW

An herbicide may be defined as any chemical agent that kills or greatly inhibits plant growth. Herbicide mode of action is the chemical interaction that interrupts a biological process necessary for plant growth and development. Herbicides injure and kill plants by interfering with the normal function of one or more of the metabolic processes vital to the life of the plant. It is generally accepted that the phytotoxicity caused by many herbicides is due to their adverse effects on normal enzyme activity. Some plants possess the means to detoxify certain herbicides and are not killed by these chemicals.

Susceptible weeds either cannot metabolize the herbicide or metabolize it too slowly for detoxification. For example, the half-life of imazaquin in soybeans is three days, whereas in cocklebur it is 30 days.

Herbicides traditionally have been discovered by screening chemical compounds in a series of increasingly specific tests. Compounds are first tested for activity against a spectrum of weeds and lack of activity against targeted crops. To be commercially successful, herbicides must have potent biological activity against a broad spectrum of weeds and, at the same time, be non-toxic to crop plants. Selective toxicity of herbicides to weeds but not to crops is one of the most difficult properties to achieve, as might be expected, from the biological relatedness of weeds and crops [83].

Both crops and weeds are naturally resistant to many selective herbicides. Herbicide resistance can be due to three basic mechanisms: resistance at the site of action, metabolic detoxification, and prevention of the herbicide from reaching the site of action.

Resistance at the site of action generally is due to an alteration in the herbicide target site (generally a protein) that prevents the herbicide from binding to the site and inhibiting the vital process mediated by the target site. Metabolic detoxification generally means that the plant can degrade the herbicide faster than the herbicide can cause irreversible damage to the plant. For an herbicide to be effective, it must reach its molecular target site. Thus, blockage of movement to the site of action can occur at the plant surface, just inside the plant or in vascular tissue where the herbicide can be rendered metabolically inactive. For a number of important classes of herbicides selectivity results from a unique or enhanced metabolic detoxification of the herbicide by the crop plant, but not by the weed.

Following their entry into plants, herbicide ions and molecules are subjected immediately to the influence of interactions with plant constituents and processes (chemical, biochemical and physical) that may affect their translocation or phytotoxicity. Once absorbed by a plant cell, an herbicide may react with chemicals present in the cell to form complexes, or conjugates, that are insoluble or non-translocatable. Such complexes immobilize the herbicide in the cell. Some herbicides are used directly on weed foliage while others are applied to the soil and depend on uptake by the weed. Certain herbicides are active in the soil and available for plant uptake for a considerable period of time while others exhibit little or no soil residual activity.

The primary mechanisms that regulate soil activity are (1) the feeding of soil bacteria on herbicides and (2) the binding of herbicides to soil. Breakdown is generally fastest in warm, moist, light textured, low pH soils and slowest in cold, dry, heavy, high pH soils.

Bacteria are the most numerous of the free living micro-organisms in the soil. Their populations in soils are unevenly distributed, commonly clustered in colonies of a few to many thousands of individuals. The colonies are scattered along the walls of pores and channels in the soil, as well as over the surface of soil particles. So enormous is the total amount of live and dead micro-organisms in soil, that soil protein is composed largely of their remains. Bacteria are the smallest living organisms, being about one micron in length.

Soil bacteria must have food for energy and growth. Organic compounds of the soil provide this food supply. If an organic substance is applied to the soil, micro-organisms immediately attack it. Those that can utilize the new food supply flourish and increase in number. In effect, this hastens decomposition of that organic substance. Organic compounds are those chemicals that contain one or more atoms of the element carbon. Carbon atoms, alone or in a chain, are able to bond with as many as four other atoms at the same time. These atoms may be carbon or other elements. Organic herbicides, in general, are composed of molecular combinations of carbon, hydrogen, oxygen, nitrogen, sulfur, phosphorus and the halogens (fluorine, chlorine, iodine and bromine).

Bacteria are capable of breaking apart the molecules of organic herbicides, resulting in their deactivation. The organisms rapidly multiply as they utilize the herbicide as an energy source. Herbicidal decomposition continues indefinitely until it is reduced completely to carbon dioxide, water and basic elements. As the food source is depleted, the large microbial population dies back.

Adsorption in soils is the process whereby ions and molecules are bonded to the surface of soil colloids due to the electrical attraction between themselves and the colloidal particles, a process similar to the attraction of iron filings to a magnet. Adsorbed herbicides are in a passive state – unavailable to biological and chemical processes.

HERBICIDE USE IN SOYBEANS

U.S. soybean growers began to use herbicides for weed control in the late 1950's. By 1966, only about 30% of the nation's soybean acreage was treated with herbicides. However, herbicide usage in soybeans grew rapidly in the late 1960's and early 1970's and reached more than 90% acreage treated by 1982. Since 1982, soybean growers have consistently treated more than 95% of the nation's soybean acreage with herbicides to control weeds. Figure 5 charts the percentage of soybean acreage treated with herbicides 1959-1998.

Significant shifts have occurred in the specific herbicides used in soybeans. Table 5 shows estimates of the percent of the nation's soybean acreage treated with specific herbicides for selected years 1966-1995. Because national surveys were not conducted between 1982 and 1990, soybean herbicide use data collected in Illinois surveys 1982-1992 are presented in Table 6. Tables 5 and 6 show the rise and fall in usage for the major herbicides used in soybeans 1966-1995. Because the individual herbicides are used at different rates per acre, a shift in usage among herbicides can lead to overall changes in the amounts of herbicide active ingredients being applied. Figure 6 charts the national usage of soybean herbicides in terms of pounds applied. As can be seen, a large decrease in pounds applied occurred in the late 1980's as growers switched to lower use rate herbicides.

The usage trends among the major herbicides used in soybeans are described in the following parts of this section. As can be seen, soybean growers have changed their use of specific herbicides. These shifts are due to differences in effectiveness in controlling weeds, amount of crop injury to soybeans and cost of alternative herbicides.

Historical Overview

The Early 1960's

The main herbicides recommended for weed control in soybeans in the late 1950's were 2,4-D, dinoseb, naptalam, chlorpropham, and CDAA. Applications of these chemicals controlled many weed species, but occasionally severely injured soybeans and provided erratic weed control. The limited acceptance of naptalam and chlorpropham was due in part to the narrow spectrum of weeds controlled relative to product cost. Following its introduction in the 1950's, chloramben (Amiben) dominated the soybean herbicide market during the 1960's. Chloramben applied to the soil surface before emergence of weeds in soybeans controlled annual grass weeds effectively when rainfall after application was adequate. In addition, chloramben controlled most small-seeded broadleaf weeds (ragweed, smartweed, velvetleaf) satisfactorily. Chloramben provided residual control of germinating weeds for 6 to 8 weeks. Chloramben inhibited root development of seedling weeds. Soybeans tolerated chloramben due to reduced translocation of the herbicide within the soybean plant in comparison to susceptible weeds. Chloramben lost effectiveness if prolonged dry weather followed its application. Chloramben was used as an over the row band. As a result, weeds between the rows were not controlled by chloramben, making cultivation still necessary. Chloramben's use as a broadcast application over an entire soybean acre was considered prohibitively expensive by most producers. Unfortunately, the cost of manufacturing chloramben was relatively expensive and there was little opportunity for a price reduction in the marketed product. As broadcast applications of other herbicides became popular, use of chloramben declined.

Soil Applied Herbicides

Trifluralin (Treflan) became available in 1963 for use as a preplant incorporated treatment for control of weeds in soybeans, applied to bare soil before planting and incorporated into the top few inches of soil with a disk. The acceptance of this treatment marked the beginning of an era in which farmer opinion shifted to allow not only a more expensive broadcast application, but also an extra tillage application for the purpose of incorporating the herbicide. Trifluralin readily volatilizes from soil surfaces and performs best when incorporated into the soil soon after application because of the loss of herbicidal activity when left on the soil surface. The universally recognized plant damage caused by trifluralin is a swelling of the root tip where cell division is disrupted. Trifluralin is root absorbed and absorbed by the shoots of seedlings as the shoots grow and push through the soil toward the soil surface. Affected weeds are prevented from emerging through the soil surface Plant death primarily results from the lack of a functional root system, leading to a lack of water flowing to the aerial parts of the plant. Trifluralin inhibits cell division by binding to tubulin and blocking the formation of spindles or microtubules necessary for cell wall formation.

Microorganisms are believed to contribute to the degradation and disappearance of trifluralin from soil.

Trifluralin is transformed by soybeans into a number of non-toxic metabolites. Soybean tolerance to trifluralin is also due to the selective placement of the chemical above the root zone of soybeans. Trifluralin is strongly absorbed on soil and shows negligible leaching. Trifluralin has been shown on occasion to have adverse effects on soybeans. Trifluralin has been shown to inhibit soybean root growth.

There have also been reports of trifluralin causing decreases in nodulation and yield of soybeans. Microbial degradation of trifluralin depends on environmental conditions – it occurs faster when the weather is hotter and wetter. Under normal conditions trifluralin provides residual control of germinating weeds throughout the soybean growing season. Residues of trifluralin may persist into the following season enough to injure certain rotational crops, such as corn and sugarbeets.

Trifluralin controls annual grasses (foxtail, crabgrass, johnsongrass) and certain broadleaf weeds (pigweeds, lambsquarters). A factor limiting the intitial acceptance of trifluralin was the lack of an economical broadleaf herbicide that could be applied with trifluralin. Since trifluralin only controls grasses and certain small-seeded broadleaf weeds, the result was often a nearly pure stand of uncontrolled broadleaf species such as jimsonweed, cocklebur, or smartweed. Soybeans often required a timely cultivation to control weeds that escaped early season control measures.

The introduction of metribuzin (Sencor, Lexone) became the catalyst for the increased acceptance of pre-plant incorporated treatments in soybeans. Research demonstrated that metribuzin provided greater than 95 percent control of the broadleaf weeds velvetleaf and smartweed, plus 60–70% control of cocklebur. A tank mix combination of trifluralin and metribuzin was evaluated extensively during 1972 – 1973 and demonstrated that the combination of metribuzin and trifluralin provided greater than 88% control of foxtail, pigweeds, velvetleaf, smartweed and morningglory, and 70% control of cocklebur.

Metribuzin inhibits photosynthesis by binding to a protein which results in the blocking of electron transport. This stops carbon dioxide assimilation. The most important detoxification route for metribuzin in soybeans is glucose conjugation to non-toxic metabolites. Soybeans degrade metribuzin to these nontoxic metabolites far more rapidly than susceptible weed species.

Differential responses of soybean cultivars to metribuzin soil applications were reported in numerous studies in the 1970's. At the rates required for effective pre-emergence weed control, soybean tolerance to metribuzin is narrow. Some soybean cultivars were determined to be more tolerant of metribuzin as a result of more rapid metabolism to nontoxic metabolites. Some soybean cultivars are so sensitive that metribuzin cannot be used safely. This very narrow margin of soybean selectivity is based on soil pH (more damage at higher pH), soil organic matter (more damage at lower organic content), rainfall (more damage with more rainfall), and rate (more damage at higher rates). Unacceptable injury from metribuzin occurred in low areas where water stood for extended periods.

In the Mississippi Delta region, successful production of soybeans requires cultivars with at least a moderate level of resistance to the plant disease phytophthora rot and to the foliar diseases bacterial pustule and target spot. The cultivar Tracy has these characteristics and ranked as the top yielding cultivar on clay soils in Mississippi.

However, tests showed that Tracy was extremely sensitive to metribuzin injury. A greenhouse technique was utilized for screening Tracy seedlings for metribuzin tolerance, after field observations had indicated that a small percentage of Tracy plants might be tolerant. Plants were selected that survived a metribuzin rate twice as high as that required to kill normal Tracy plants. Surviving seedlings were transplanted and grown to maturity. A new cultivar called Tracy-M was registered. Tracy-M is tolerant to metribuzin and is similar to Tracy in yield and disease resistance. Seed was distributed in 1979 for increase in Mississippi, Alabama,

Arkansas and Louisiana. These plants appear to be tolerant owing to a detoxification mechanism resulting in inactivation of metribuzin.

The herbicide alachlor (Lasso) was registered for use in soybeans in 1969. Alachlor controls many annual grasses and some small-seeded broadleaf weeds. As metolachlor (Dual) was introduced, it was in direct competition with alachlor and obtained some of that market. The mechanisms of action of alachlor and metolachlor are not well understood. Alachlor generally provides six to ten weeks of weed control while metolachlor generally provides ten to fourteen weeks of weed control. The effectiveness of alachlor and metolachlor is improved when they are incorporated into the soil mechanically.

The dinitroanaline herbicide pendimethalin (Prowl) was registered for soybeans in 1976. Pendimethalin is soil-applied before planting and is absorbed by plant roots. Pendimethalin is immobile, being strongly bound to organic matter and clay. Typical half-life in the field is 44 days. Highly susceptible annual grasses and broadleaves usually fail to emerge. Pendimethalin does not require mechanical incorporation to be effective. Pendimethalin is less volatile than trifluralin and need not be incorporated for seven days following its application. Thus, early rains can often be relied upon for its incorporation.

Postemergence Herbicides

Postemergence herbicides are applied to emerged weeds. Postemergence herbicide programs allow growers to adjust the herbicide rate and application date to weed growth depending on the severity of the weed problem and to target specific weed species.

Certain broadleaf weeds (such as cocklebur) which were not controlled adequately with the available preplant or preemergence herbicide treatments became greater problems in the early 1980's. Soybeans often required a timely cultivation to control weeds that escaped the early season control measures. Several herbicides were introduced in the 1970's and 1980's that provided selective control of several of these troublesome weeds when applied postemergence.

Research demonstrated that postemergence application of bentazon (Basagran) over the top of weeds and soybeans selectively controlled many broadleaf weed species without reducing soybean yields. Bentazon controls ragweed, smartweed, velvetleaf and common cocklebur. Bentazon proved erratic in controlling pigweeds and lambsquarters. Weed tolerance to bentazon increases rapidly as they grow larger. Research demonstrated that a preemergence treatment of metribuzin and a postemergence application of bentazon provided 92–99% control of common cocklebur.

Bentazon inhibits photosynthesis by binding to a protein which blocks electron transport. In soybeans bentazon is rapidly metabolized following glucose conjugation. Bentazon has little to no soil residual activity. However, bentazon applications may result in foliar bronzing of soybeans.

Soybean plants usually will outgrow the leaf burn provided they are not subjected to additional stress and have ample time to recover. The use of trifluralin for grass control followed by postemergence applications of bentazon for broadleaf weed control proved popular in the 1980's.

Research demonstrated that acifluorfen (Blazer) controls a number of weeds in soybeans when applied postemergence: cocklebur, pigweed and morningglory. Soybean injury from postemergence application of acifluorfen is exhibited as leaf crinkling and red speckling or bronzing. Soybeans normally recover 10 to 14 days after application and continue growing at a normal rate.

The primary target site for acifluorfen appears to be Protox, an enzyme of chlorophyll. Soybean tolerance appears to be due to rapid cleavage of the ether bond followed by further metabolism. Metabolism appears to be much slower in susceptible weed species. Acifluorfen is readily absorbed by leaves and is considered to be a contact herbicide since little translocation occurs and leaf injury is usually observed rather quickly. The best weed control with acifluorfen is obtained when weeds are small and actively growing. A potential problem with postemergence mixtures deals with the timing of applications in instances where plant size between different weed species is highly variable. This is a concern in tank mixes of acifluorfen and bentazon in which the window of application of each herbicide is short and specific for the size or stage of growth of weeds.

Sethoxydim (Poast) and fluazifop (Fusilade) are postemergence herbicides that selectively control grass species in broadleaf crops, such as soybeans. Fluazifop and sethoxydim became available for use in soybeans in the early 1980's.

Researchers have reported excellent control of may grass species with fluazifop and sethoxydim at very low rates. Fluazifop and sethoxydim are absorbed rapidly into leaves and are rainfast within one to two hours. Grasses must be growing actively for these herbicides to be effective. Sethoxydim and fluazifop have short field half-lives of 5-15 days. Fluazifop occasionally controls or suppresses grass weeds germinating after application.

Sethoxydim and fluazifop inhibit the enzyme ACCase, that inhibits the production of membranes required for cell growth. In soybeans, transformation of sethoxydim and fluazifop to metabolites is very rapid. The slow adoption of the postemergence grass herbicides (such as sethoxydim and fluazifop) is primarily attributed to high product cost, and the risk associated with late season treatments.

One limitation on the use of postemergence herbicides in soybeans is the inability to tank mix broadleaf and grass herbicides. A tank mix of bentazon or acifluorfen with sethoxydim or fluazifop produces an antagonism which reduces the efficacy of the grass herbicides: the burning of the grasses by the broadleaf materials reduces the uptake and transformation of the grass herbicide by the weeds.

Additional postemergence grass control herbicides were introduced into the soybean market in the 1990's: quizalofop, clethodim and fenoxaprop. Additional broadleaf control herbicides introduced for soybeans included fomesafen, lactofen, flumiclorac and sulfentrazone.

Flumetsulam was introduced in 1993 as a component in mixtures with trifluralin and metolachlor to provide broadleaf control. Clomazone was commercialized in 1985 and was targeted for the control of velvetleaf. Clomazone also controls several other broadleaf weeds and several seedling grasses. Availability of effective postemergence herbicides allowed some soybean producers to switch entirely to postemergence weed control programs. In some cases, postemergence applications were made to soybeans because of inadequate performance of soil applied materials during years with dry soil conditions.

Sulfonylurea/Imidazolinone Herbicides

In the mid and late 1980's, several herbicides from the sulfonylurea and imidazolinone chemical classes were introduced for use in soybeans: imazethapyr (Pursuit), imazaquin (Scepter) and chlorimuron (Classic). The sulfonylureas and imidazolinones are similar in their mode of action. They are absorbed readily by roots and foliage of plants.

They are transported throughout the plant with concentration in merismatic tissues (buds), a primary site of DNA, and amino acid synthesis. Herbicide activity by the sulfonylureas and imidazolinones results from inhibition of the enzyme acetohydroxyacid (AHAS, acetolactate synthase, ALS), which stops the synthesis of three essential amino acids – valine, leucine and isoleucine. This causes a disruption in protein synthesis resulting in an interference of DNA synthesis and rapid cessation of growth. Metabolic inactivation by the soybean plant serves as the basis of selectivity.

Soybean plants rapidly metabolize imazaquin to a wide variety of compounds that either are degraded or incorporated into naturally occurring plant constituents. Imazethapyr is metabolized to non-toxic forms in soybeans by hydroxylation followed by conjugation to glucose. The imidazolinones and sulfonylureas are used at low rates (.004-.125 LB AI per acre) and can be applied in a variety of ways. Imazaquin and imazethapyr can be applied preplant, preemergence or postemergence. Chlorimuron is marketed for postemergence herbicide applications. Chlorimuron and imazaquin are primarily targeted at broadleaf weeds in soybeans.

Research demonstrated that soil applied imazaquin provided weed control superior to metribuzin on cocklebur (93 vs. 74%), giant ragweed (93 vs. 63%) and velvetleaf (91 vs. 80%). Research demonstrated that imazethapyr controlled grass and broadleaf weeds effectively.

Applications of imazethapyr postemergence following pendimethalin applied preplant provided effective control of giant foxtail, pigweed, velvetleaf, cocklebur and lambsquarters. This combination provided improved control over a broader spectrum of weeds than a combination of bentazon plus acifluorfen following pendimethalin.

Imazaquin and imazethapyr are used widely in soybeans due to their weed control advantages and the flexibility to apply them preemergence, preplant incorporated or postemergence. Imazethapyr exhibits soil residual weed control activity, yet controls a broader spectrum of weeds when applied postemergence. Imazaquin evaluations in 1985 indicated greater than 90% control of cocklebur, pigweed, lambsquarters, smartweed, ragweed and giant foxtail.

Research indicated that postemergence applications of imazethapyr and imazaquin discolored and restricted margins in young, expanding soybean leaves. Also, imazethapyr treated soybeans were significantly shorter compared to untreated, handweeded soybeans.

However, it was reported that this condition did not affect soybean yield. Because of its limited spectrum on grasses, imazaquin was labeled for tank mixing with such soil-applied grass herbicides as pendimethalin and trifluralin. However, research demonstrated that mixtures of imazaquin with the postemergence herbicides sethoxydim and fluazifop had an antagonistic effect and reduced grass control.

The sulfonylureas and imidazalones remain active in the soil after application for season-long weed control. They are degraded in the soil via environmental routes, such as microbial activity and plant uptake.

The greatest contribution to breakdown by microorganisms occurs in warm, moist soils. Conversely, cool temperatures and either very wet or very dry soil conditions lead to a slower microbial degradation of the sulfonylureas and imidazalones. Since sufficient concentrations of the imidazalones or sulfonylureas can remain in soil to injure certain sensitive rotational crops (such as corn, sorghum, cotton and rice) restrictions have been placed on the crops that can follow soybeans. Susceptible crops that are seeded in years following imazethapyr application can be damaged severely by imazethapyr residues in the soil.

One limitation to the use of ALS herbicides is the development of resistant weed populations. Weeds that have evolved resistance to ALS inhibitors have evolved altered ALS that is resistant to the herbicide. Common waterhemp is a widespread problem in Midwest soybean production because of resistance to ALS-inhibiting herbicides. In Kansas resistance to ALS-inhibiting herbicides has been confirmed in kochia, Russian thistle, common waterhemp, Palmer amaranth, common cocklebur, shattercane and common sunflower.

The occurrence of herbicide resistant weed biotypes in the Midwest increased in the mid 1990's. Biotypes of common and tall waterhemp and kochia demonstrating resistance to ALS-inhibiting herbicides have been reported. The prevalence of herbicide resistant weeds has led to postemergence mixtures of active ingredients for broad range weed control.

Early field testing in the U.S. indicated that imazaquin provided excellent control of common cocklebur, that was not controlled adequately by the standard treatments available at the time. Tests in Mississippi demonstrated that imazaquin controlled sicklepod. In 1985, the EPA permitted growers in five southern states to use imazaquin under a Section 18 Emergency Exemption Clause for control of sicklepod since there were no registered herbicides that controlled this weed in soybeans. In 1986, imazaquin was registered for use in soybeans in the U.S., where it is sold under the trade name of Scepter.

In 1982, field testing of imazethapyr was begun in U.S. soybeans. Initial results demonstrated that imazethapyr controlled a wide spectrum of weeds, including velvetleaf, pigweeds, nightshades and foxtails. University researchers in Minnesota demonstrated that imazethapyr was highly effective in controlling Jerusalem artichoke. Because there were no labeled herbicides that controlled this weed in soybeans, EPA permitted Minnesota growers to use imazethapyr under a Section 18 Emergency

Exemption Clause in 1987 and 1988. In 1989, imazethapyr was registered for use in soybeans in the U.S. Pendimethalin's (1 rowl) use increased from 3 to 23% acreage treated as a result of its availability in formulated mixtures with imazaquin and imazethapyr, and its use for surface application with reduced tillage. The soybean acreage treated with trifluralin, which requires preplant incorporation, declined from 61% treated in 1982 to 50% in 1990.

In a reduced tillage system, it is not possible to incorporate herbicides into the soil as a preplant treatment.

The use of metribuzin decreased from 1982 to 1990 as soybean producers used other herbicides in an effort to reduce crop injury and broaden the spectrum of weed control. Herbicides used as metribuzin replacements included chlorimuron and imazaquin. Use of these herbicides resulted in a 50% decline in metribuzin use within one year of their introduction in 1987.

An aggressive imazaquin marketing campaign led to its widespread use in 1988, a drought year in the Midwest. The subsequent widespread carryover injury to corn in 1989 resulted in a change in the label and elimination of imazaquin use in the northern two-thirds of the Corn Belt.

The introduction of imazethapyr in 1989 resulted in a decline in both metribuzin and imazaquin use.

Because chlorimuron and thifensulfuron are broadleaf herbicides, they need to be mixed with grass herbicides for broad-spectrum postemergence weed control in STS soybeans. Because many fields contain ALS-resistant waterhemp, postemergence treatments are often tank mixed with lactofen or acifluorfen. For preemergence applications, a copack of thifensulfuron and chlorimuron plus sulfentrazone provides control of waterhemp and nightshade in addition to broadleaf weeds. This copack can be combined with a grass herbicide such as pendimethalin.

Burndown Herbicides

Soybeans often follow corn, which has a large amount of crop residue. Leaving residues intact or mixing them into the soil surface is highly desirable for the purpose of controlling erosion.

Prior to the widespread use of herbicides; surface trash was unacceptable because it clogged the cultivator that was used for weed control.

Conservation tillage practices, including no-till, that greatly reduced preplant tillage for weed control increased reliance on herbicides for weed control. In the 1970's the herbicide paraquat (Gramoxone) was introduced to kill existing vegetation at the time of crop planting. After spraying paraquat, all the plants were killed and completely dead, almost as if they had been burned with fire.

For this reason, the term "burndown" was coined. Another commonly-used burndown herbicide is glyphosate (Roundup). It also provides broad-spectrum control of both grass and broadleaf, annual and perennial vegetation. A commonly-used burndown herbicide is 2,4-D, an inexpensive herbicide that is effective on winter annuals/perennials and annual broadleaves, but is not effective on annual grasses.

Control of existing vegetation at planting is absolutely critical to successful no-till farming. If any weeds survive, they will have a head start on the soybean plants and will be extremely difficult to control.

Summary of Usage: 1995

By the early 1990's, there were at least 70 registrations for individual herbicides or packaged herbicide mixtures for weed management in soybeans. As a result, most weeds in soybeans could be adequately controlled with the herbicides available in the early 1990's in well-planned management systems.

During the 1990's soybean growers increasingly shifted towards postemergence herbicide use. The average number of treatments per acre rose from 1.5 in 1990 to 1.7 in 1995 as it was more common for soybean growers to make both an at-plant and postemergence treatment or make two postemergence treatments.

Table below compares 1988 with 1995 in terms of the distribution of soybean acreage by the number of herbicide applications. As can be seen, there was a decline in acreage that received a single herbicide application (51 to 41%) with a concomitant rise in the number of acres that received two treatments (34 to 44%).

As can be seen in Table, a steady increase in the number of herbicide active ingredients applied to treated soybean acreage occurred between 1986 and 1995: from 1.4 to 2.7.

In 1995, 23% of the nation's soybean acreage was treated with a combination of four or more active ingredients while 28% received three active ingredients, 35% received two active ingredients and 12% was treated with just a single herbicide active ingredient.

Annual soybean yield loss to weeds in the U.S. was estimated at 17% during the 1951-60 time period. In 1992, U.S. soybean yield losses to weeds was estimated at 7%.

The increased use of herbicides led to significant reductions in the number of cultivations of U.S. soybean acreage. In 1994, only 43% of U.S. soybean acreage received any cultivation during the growing season. The average cultivated acre was cultivated one time during the season in 1994.

Table : Herbicide Use on Illinois Soybean Acreage (1982-1992)

	% Acres Treated		Active Ingredient		
	1982	**1985**	**1988**	**1990**	**1992**
Acifluorfen	1	6	9	8	11
Alachlor	25	15	11	13	9
Bentazon	20	26	30	32	24
Chloramben	4	3	1	1	-
Chlorimuron	9	17	20		
Imazaquin	27	12	15		
Imazethapyr	11	28			
Metolachlor	10	14	10	8	5
Metribuzin	51	46	23	16	13
Pendimethalin	3	7	15	24	32
Trifluralin	61	48	43	39	27

TABLE : Soybean Area by Number of Herbicide Applications (%)

	1988	1995
No Applications	4	3
One Application	51	41
Two Applications	34	44
Three or more Applications	11	12

TABLE : Number of Herbicide Active Ingredients

Applied per Treated Soybean Acre

Year	Number
1986	1.4
1988	1.6
1990	2.3
1991	2.3
1992	2.4
1993	2.5
1994	2.7
1995	2.7

12

Transgenic Herbicide Tolerant Soybeans

GLYPHOSATE – AN OVERVIEW

Glyphosate was first introduced as an herbicide under the trade name of Roundup by Monsanto in 1974. Glyphosate is a nonselective broad spectrum herbicide that normally cannot be applied over crops without severe plant injury. Some plant species are somewhat tolerant to glyphosate, probably because of uptake or translocation mechanisms, but no crop has sufficient tolerance for glyphosate to be used directly on the crop without damage. Glyphosate is a systemic herbicide that is transported rapidly from the foliar tissue to the metabolically active regions of shoots and root tips]. Within these tissues, glyphosate inhibits the biosynthesis of aromatic amino acids by inhibiting the enzyme EPSPS, an enzyme essential for amino acid synthesis.

This pathway for aromatic amino acid biosynthesis is required in plants, not only for the formation of the aromatic amino acids, but also for a number of vitamins, phytohormones and lignins. It is estimated that 40-60% of the carbon fixed by plants is fixed through this pathway. Glyphosate inhibits EPSPS, thus preventing the plant from making the aromatic amino acids essential for the synthesis of proteins and some secondary metabolites.

EPSPS is the only physiological target of glyphosate in plants. EPSPS is present in all plants, bacteria and fungi, but not in animals. In plants, EPSPS is localized in the chloroplasts or plastids. Only glyphosate-based herbicides have this biochemical mode of action. No other class of commercial herbicides are known to target this enzyme.

Glyphosate is bound tightly to soil particles and is metabolized readily by soil micro-organisms into plant nutrients. As a result, glyphosate exhibits no residual soil activity. The initial inactivation of glyphosate in soils is believed to result primarily from the formation of insoluble complexes with aluminum, iron and calcium ions bound to the soil. Degradation by soil microflora is the predominant route by which glyphosate is metabolized in the environment. Studies on the degradation of glyphosate in the soil indicate the formation of ammonia, phosphates, and carbon.

Section 6 includes a description of the development of transgenic soybean plants that tolerate glyphosate applications following the insertion of a gene from a soil bacteria that provides the soybean plant with a version of the enzyme EPSPS to which glyphosate cannot bind. The transgenic soybeans were introduced to U.S. growers in 1996. Glyphosate had been used on 20% of the nation's soybean acreage in 1995 – the year before the introduction of the transgenic varieties.

Prior to the development to the transgenic cultivars, glyphosate could not be used directly over the top of soybean plants. Glyphosate's use in soybeans prior to 1996 was confined to use as a non-selective "burndown" of weeds prior to soybean planting, spot spraying and in a recirculating sprayer.

High clearance sprayers, known as "bean buggies," were developed, allowing operators to spot spray glyphosate directly on individual weed escapes in soybean fields, avoiding the soybean plants. For tall-growing weeds in soybeans, the recirculating sprayer may be used. It relies on a weed-to-crop height differential by which tall weeds can be controlled selectively with a non-selective herbicide in a short-statured crop. The basic concept of the sprayer is to direct the spray above the crop. A catch basin over the row for each spray nozzle catches any spray that is not intercepted by tall weeds and recirculates it through the sprayer. Weeds must be at least 16 cm taller than the soybeans. Research demonstrated that use of glyphosate in the recirculating sprayer to control johnsongrass greatly increased soybean yields.

PERFORMANCE OF ROUNDUP READY SOYBEANS

The use of glyphosate over-the-top of Roundup Ready soybeans was researched extensively prior to and immediately after their introduction in 1996. Roundup Ready soybeans were evaluated under an EUP in 1995 in 14 states in the Midwest and Mid Atlantic. A single application of glyphosate at 0.63-0.84 kg/ha provided annual weed control throughout the entire growing season.

In a 1997 experiment, all application timings of glyphosate (at weed heights of 7.5, 15, 23 and 36 cm) provided season-long control of giant foxtail, common lambsquarters and common cocklebur. The 7.5 cm timing was the only treatment to provide better than 90% control of smartweed. (All other glyphosate treatments provided 78-89% control.) All glyphosate applications provided yields equal to the weed-free comparison. All comparison treatments yielded significantly lower than weed-free comparison. Glyphosate applications provided excellent control of ALS-resistant waterhemp when applied postemergence over glyphosate tolerant soybeans. In Missouri, common waterhemp resistant to the sulfonylureas and imidazolinones had become the principal weed problem in soybeans with over 900,000 acres of soybeans infested. In 1996 experiments, imazaquin, imazethapyr and chlorimuron provided 15-25% control while.75 lb. AI of glyphosate applied over glyphosate tolerant soybeans provided 98% control.

Experiments in 1995 and 1996 in Illinois demonstrated that the use of a tank mix partner or a preplant herbicide followed by a single glyphosate application improved weed control compared with a single application of glyphosate. However, the highest levels of weed control were always obtained with two sequential applications of glyphosate.

Experiments compared the yield impact of glyphosate and alternative herbicide treatments. At seven locations in 1996, yield averaged across seven locations was higher in the glyphosate treated acres in comparison with those treated with imazethapyr, acifluorfen, bentazon and sethoxydim.

Glyphosate's lack of residual activity raises concerns of the optimal time to apply glyphosate and whether sequential applications will be needed. Experiments demonstrated that glyphosate applied at 5 or 8 weeks after soybean planting provided season-long control of almost all weed species.

When glyphosate is applied early (when weeds are 7.5 cm), a late flush of a weed species such as velvetleaf is not controlled, and yield can be reduced. In a 1998 experiment, yield loss due to weeds germinating after glyphosate application was observed with the one-pass, two to four inch weed height application.

Proper timing of glyphosate application is necessary because applications too early following soybean planting (10 days) allow significant reinfestation of weeds late in season. Also, delaying treatment past 28 days allows an unacceptable level of weed competition reducing soybean yield due to early season weed pressure.

Experiments were established at 15 sites in 1997 and 12 sites in 1998 across the north central and northeast U.S. to evaluate the time of application of Roundup (based on weed size) on soybean yield in a Roundup Ready system. When weeds were treated at the 7.6, 15.2 or 22.8 cm stage, the Roundup plots were among the highest yielding treatments at all sites. When weeds were treated at the 30.5 cm stage, yield reductions occurred at 6 of the 15 sites.

In southern states, rainfall distribution throughout the growing season results in several flushes of weeds. In Mississippi, with a long growing season and a more difficult weed complex, a minimum of two applications of Roundup is necessary.

Due to the extended emergence period of broadleaf signalgrass and large crabgrass, at least two applications of Roundup are generally required for season-long control of these weeds.

The University of Arkansas recommendation is to apply one pint of Roundup 10 to 14 days after the soybeans and weeds emerge and to repeat the treatment in 7 to 10 days. This treatment schedule consistently has produced equal control across all weed species.

As with any other herbicide, weed species vary in their tolerance to glyphosate. Highly susceptible weeds, such as giant foxtail, can be controlled with lower rates than required to control weeds with a high level of tolerance, such as velvetleaf. The typical use rates of Roundup (24 to 32 ounces per acre) were chosen since they are the lowest rates that will provide consistent control of most major weeds found in soybean fields.

HERBICIDE RATINGS

Each year, state extension services release weed control guides for field crops which include soybeans. The guides advise growers on the expected performance of available herbicide treatments to control specific weed species. The guides also include a rating as to the likely injury to soybeans that may result from the treatment. The efficacy rating for a particular herbicide treatment to control a particular weed species is developed by extension specialists based on experimental work and the experiences of farmers in the state. Generally, the rating tables include five broad categories of control effectiveness: none, poor, fair, good and excellent. These ratings are associated with a range of percent control of the weed species.

For example, a common range for a "good" rating is 80 to 90% control. The rating tables group herbicide treatments according to timing: preplant incorporated (ppi), pre-emergence (pre), or postemergence (post). The most extensive set of soybean herbicide treatments rated in a university weed control guide is issued annually by Michigan State University (MSU). The MSU guide rates the effectiveness of 182 herbicide treatments in controlling 24 different weed species. The MSU guide rates the effectiveness of 47 treatments that contain a single active ingredient, 103 treatments that contain two active ingredients and 33 treatments that contain 3 active ingredients. The MSU guide rates the potential for soybean crop injury from the treatments as follows: 47 treatments have a "minimal risk of crop injury," for 53 treatments "crop injury can occur under certain conditions," and for 66 treatments "severe crop injury can occur."

Roundup (glyphosate), used over Roundup Ready soybeans, is one of the 182 treatments rated by MSU. MSU rates the effectiveness of Roundup as "good" or "excellent" on 23 of the 24 weed species. Roundup receives a rating of "fair" for yellow nutsedge control. The 23 good/excellent ratings is the highest total of good/excellent ratings for the 182 herbicide treatments.

In addition, the Roundup treatment is assigned a rating of minimal risk of crop injury.

The next highest total of good/excellent ratings (18) is for a preemergence application of either dimethenamid, or alachlor in combination with metribuzin plus chlorimuron. This combination is also assigned one fair rating, 3 poor ratings, and 2 ratings of no control. This treatment is assigned a crop response rating of "severe crop injury can occur." The most effective postemergence treatment in terms of the number of good/excellent ratings (17) is the prepackage mixture of bentazon and acifluorfen plus sethoxydim (this combination is rated as two separate treatments in the MSU table).

This treatment also receives a rating of "severe crop injury." Of the 47 herbicide treatments that have a rating of "minimal" crop injury, the most highly rated alternative to the Roundup treatment is a preemergence combination of either dimethenamid, alachlor or metolachlor in combination with imazaquin. This treatment is assigned 16 ratings of good or excellent. The most effective postemergence herbicide treatment with a minimal crop injury risk rating is the combination of chlorimuron and thifensulfuron for use with STS soybeans, which is assigned 11 good and excellent ratings with 12 ratings of "none" – mostly for annual grasses.

A combination not rated by MSU because it would require two trips across the field is the use of fluazifop for grass control followed by a chlorimuron/thifensulfuron treatment for broadleaf weed control. These products are assigned a rating of minimal crop injury. Two passes are required for this treatment since the treatment is not labeled for tank mixture due to an antagonistic interaction which reduces effectiveness on grasses. This treatment receives 20 good/excellent ratings with three ratings of "none" due to a lack of control of certain perennials.

The MSU weed control guide for soybeans also includes a table advising growers of the maximum broadleaf weed heights for consistent control. Table below shows these maximum weed heights for five of the postemergence treatments for 11 broadleaf weed species. As can be seen, for all weed species glyphosate can be applied up to a maximum weed height of at least four inches.

The combination of chlorimuron plus thifensulfuron can also be used on weeds up to four inches. However, imazethapyr will only control lambsquarters if they are less than one inch tall while the combination of acifluorfen plus bentazon can be used to control redroot pigweed at a maximum height of only two inches.

Because of the potential for injury to sensitive crops, restrictions on the label identify the minimum interval required before certain crops can be grown in rotation with soybeans treated with certain herbicide treatments. Table 13 shows these minimum intervals for four herbicide treatments. Because of its lack of soil residual activity, there are no rotation restrictions on the glyphosate label for any crop. As can be seen in Table 13, a very sensitive crop such as sugarbeets cannot be planted into a soybean field for 26 to 40 months following the applications of chlorimuron/metribuzin, imazaquin, imazethapyr or thifensulfuron/chlorimuron. The common soybean rotational crop is corn, that can be planted eight to ten months following the application of chlorimuron, metribuzin, imazethapyr or thifensulfuron in soybeans.

Adoption Impacts: 1995 – 1998

As can be seen, glyphosate usage increased to 46% of acreage treated in 1998, from 20% in 1995, while most other active ingredients recorded declined in acreage treated.

The impacts of the adoption of Roundup Ready soybeans and the associated use of glyphosate are presented in the next sections by comparing herbicide use in 1995, the year before their introduction, with 1998, the latest year for which herbicide use data are available. The comparison is based on USDA surveys of soybean growers in the same thirteen states in both 1995 and 1998. These thirteen states represent approximately 80% of U.S. soybean production. Consequently, the results determined from the thirteen state comparison are assumed to equal 80% of the national impacts. Other states that were included in either the 1995 or 1998 surveys (but not both) were excluded from the calculations since comparisons between the two years could not be made.

Herbicide Costs

Roundup Ready soybean weed control programs have been priced to be competitive with conventional programs. In addition to the cost of the herbicide, soybean growers must pay the equivalent of $6 per acre as a "technology fee" when purchasing Roundup Ready seed. Table 15 displays estimates of the costs of different weed control programs for soybeans. These estimates are exclusive of any burndown preplant herbicide treatments. As can be seen, depending on the rates used, numbers of applications and combinations of products, the cost of a conventional program can range from $14 to $25 per acre, an STS program can range from $11 to $28 per acre, while a Roundup Ready program can range from $16 to $32 per acre (including the technology fee).

Potential savings depend on which program growers choose. Growers tailor weed control programs to fit particular weed pressure in each field, choosing herbicides and combinations of herbicides to control particular weed species. There are no data on the specific weed control programs growers are using.

The introduction of the competitively-priced Roundup Ready weed control systems led to reductions in the prices farmers paid for competitive herbicides. Table 16 charts the per pound cost of selected herbicide active ingredients from 1990 to 1998. As can be seen, the costs of chlorimuron and imazethapyr were reduced significantly (40-50%) in 1997 and 1998.

The price of glyphosate was also reduced in 1998 by 22%. The result of lower priced Roundup Ready treatments in comparison with competitive herbicides and the lowering of the price for key herbicides including glyphosate meant that soybean growers spent significantly less on herbicides in 1998 than in 1995. Table 17 displays estimates of aggregate expenditures on soybean herbicides in the thirteen states in the USDA surveys for 1995 and 1998. These aggregate expenditure estimates have been calculated by multiplying the USDA survey estimates of pounds of each active ingredient applied times an estimated price per pound of each active ingredient.

As can be seen, the aggregate expenditure amount for the thirteen states declined from $1.5 billion in 1995 to $1.2 billion in 1998. Assuming that the thirteen states represent 80% of national production implies a reduction in herbicide expenditures from $1.86 billion in 1995 to $1.48 billion in 1998, representing a reduction of $380 million in annual herbicide expenditures by U.S. soybean growers. In addition to the cost of the herbicide glyphosate, soybean growers who purchase Roundup Ready seed paid the equivalent of an extra $6 per acre in a technology fee. As shown in Table 1, 37% of the soybean acres, or 27 million acres, were planted to the Roundup Ready soybeans in 1998, implying a total annual technology fee of $160 million. Thus, soybean growers spent $220 million less on weed control in 1998 due to lower costs after netting out the technology fee.

Soybean Yields

There are different points of view regarding the impacts of the wide-spread planting of Roundup Ready soybean on yield. On one hand, some argue that university performance data show the Roundup Ready varieties lagging behind other varieties in terms of yield. On the other hand, Monsanto has presented data that show that Roundup Ready soybeans outyielded national averages in 1998.

Determining the impact that Roundup Ready soybeans have had on yields is somewhat problematic at this stage. With such a new technology, field data on yields are scarce. However, some expectations may be developed from available research results. Two areas of research are relevant to addressing the issue of the impact that Roundup Ready soybeans have had on yields. The first type of research is variety trials, where the yield potential of conventional and Roundup Ready varieties have been compared under weed free conditions. The second type of research is weed control trials, which compare weed control strategies.

Variety trials from more than 3,000 side-by-side comparisons from forty university performance tests conducted across eight states in 1998 were summarized by a University of Wisconsin agronomist. When averaged across all tests, Roundup Ready varieties were 4% lower in yield than conventional varieties. On average, the top five Roundup Ready varieties yielded 5% less than the top five conventional varieties in 200 comparisons. The most often heard explanation for the yield lag is that the Roundup Ready gene often has not been put into the most elite lines that some companies have to offer. A more plausible explanation, say some soybean breeders, is that in the rush to get Roundup Ready lines on the market; many companies have not made enough backcrosses to capture all of the yield potential in the parent lines. In either case, the conventional wisdom says that will be corrected in a reasonably short time, and yield lags will be eliminated.

1999 University variety trial data have not been compiled as yet to compare the Roundup-Ready variety yields to those of non-Roundup-Ready varieties. Some have argued that the university trials show a "yield drag" from Roundup Ready soybeans and that national soybean yields could decline by 2–2.5% as a result of widespread planting of the Roundup Ready cultivars.

This analysis ignores potential weed control benefits of the new technology, as it is based solely on the results of the variety trials. Upon further analysis of the university trial data, Monsanto has pointed out that the notion of a yield drag is based on the overall average yields of conventional varieties in comparison with the overall average yield of the Roundup Ready varieties. Monsanto argues that such an analysis can be misleading since fewer Roundup Ready varieties were tested than the number of conventional varieties.

As a result, one poor performing variety has a large impact on the overall average. Monsanto points out that many of the elite varieties of Roundup Ready soybeans have not been included in the variety trials. Monsanto's position is that the top-selling elite Roundup Ready varieties have yields comparable to top-selling elite conventional varieties.

In soybean fields the potential yield advantage of conventional varieties may not be realized because of poor weed control. In addition, injury from herbicides may reduce the potential yield advantage of conventional varieties. In weed control trials, weed control programs are compared as to their efficacy in controlling weeds, and yields are often recorded. Several herbicides are usually included in these trials, alone or in combination, at variable rates and application timing. The purpose of these types of studies is to determine optimal rates and timing to achieve control of various weeds. In general, these tests are conducted using one variety in order to eliminate variety as a variable. Many of these studies are now conducted using Roundup Ready varieties, in order to include Roundup treatments in the studies.

The yield differences in weed control trials are due to differences in weed control and/or crop injury associated with the different herbicides tested but do not take into account the yield potential of the variety used in the study. It is difficult to generalize about the results of these weed control studies, except to note that at this point there seems to be no resounding yield advantage or disadvantage in Roundup Ready systems compared to conventional programs.

Increasingly, weed science researchers have focused on the relationship between crop injury and lower soybean yields. Injury from contact postemergence herbicides such as acifluorfen and lactofen has been observed for a number of years. Combinations of thifensulfuron and imazaquin that achieve acceptable lambsquarters control also result in unacceptable soybean injury and possible yield loss.

Soybean injury may alter the canopy such that late germinating weeds may escape control. Bentazon, acifluorfen, thifensulfuron, lactofen and clethodim treatments reduced leaf area and reduced soybean yield 209 to 215 kg/ha compared to the untreated control. Soybean injury from postemergence herbicides reduces yields. A perception that injury from postemergence herbicides may reduce soybean yield potential has been triggered by the minimal to no soybean injury observed with foliar applied glyphosate on glyphosate resistant soybeans.

Surveys conducted by Monsanto and USDA have compared yields in Roundup Ready soybeans and conventional soybeans. Monsanto reported that Roundup Ready soybeans out yielded the national average in 1998. A 4.5 bushel per acre difference was observed, based on a comparison of USDA estimated national soybean yields and Monsanto grower survey results. However, this comparison does not take into account differences in other characteristics between adopters and non-adopters of the technology.

The observed difference may be due to adoption of the technology by better managers who would obtain higher than average yields normally. The difference may also be due to adoption by growers who utilize production methods, such as narrow row spacing, that would account for higher yields.

In USDA surveys of soybean producers, yield differences were observed for growers using herbicide-tolerant varieties compared to other growers who purchased seed. Table below shows USDA survey results for 1996-1998. Statistically significant differences were observed in three production regions in 1997 and one region in 1998. All of these differences showed higher yields for growers planting herbicide-tolerant varieties. However, the USDA analysis of the survey data does not take into account any confounding factors, which may have influenced the observed differences between herbicide-tolerant varieties and conventional varieties.

Similar to the Monsanto results discussed above, the results are differences in means, which may be driven by factors other than the adoption of herbicide-tolerant varieties. In fact, further analysis indicated that only a small increase in yields, less than 1% for a 10% change in adoption, was related to adoption of herbicide tolerant varieties when other factors were considered and controlled. Results of the extended analysis showed that larger operations and more educated operators were more likely to use herbicide-tolerant soybean seed. Use of conventional tillage was higher for conventional growers, which would be expected if Roundup Ready soybeans are adopted in conservation tillage systems, as is widely believed.

Researchers in Minnesota concluded that yields in a Roundup system compared to conventional herbicide systems were equal. In a summary of weed control research published in the 1997 North Central Weed Science Society Research Report, in which Roundup-only treatments were compared to conventional programs, Roundup-only plots out-yielded conventional plots by 5.3 bushels per acre. However, some conventional treatments in those comparisons may have been tested against weeds they do not control, whereas Roundup has a broad weed spectrum. This may make Roundup Ready systems appear to be more effective in these types of studies than they would be in reality, where a grower would tailor a weed control program for the particular weed species present in the field.

Returns

Researchers have compared net returns for conventional and Roundup Ready programs, taking into account both cost and yield differences between the programs. Based on data from field trials conducted in Tennessee in 1995, 1996 and 1997, comparing several conventional and Roundup treatments on Roundup Ready varieties, the Roundup only program was the most profitable system, providing a 13% higher net return than the next best alternative. In Arkansas trials conducted during 1996 and 1997, conventional herbicide programs on nontransgenic and glyphosate tolerant soybean were compared to glyphosate programs on glyphosate-tolerant soybean. Two conventional programs had the highest net returns ($37.33 and $49.72 per acre), while only two of the Roundup Ready programs had positive returns ($14.21 and $15.99 per acre).

A 1997 trial conducted in Louisiana, comparing conventional programs with programs combining conventional herbicides with glyphosate or glyphosate alone on glyphosate-tolerant varieties, resulted in the highest yielding treatment being a combination of metribuzin at a reduced rate applied preemergence followed by Roundup. Though this treatment was more costly than a Roundup only program, only a 0.5 bushel increase in yields would be needed most years to pay for the difference. Trials were conducted in 1997 in Mississippi, using the three highest yielding Roundup Ready and conventional cultivars. Net returns were more than $60.00 per acre higher with the labeled rate glyphosate system at two of three locations compared to the labeled rate conventional system.

A study at the Leopold Center that analyzed 1998 crop survey data for Iowa concluded that returns to land and labor essentially were identical for the genetically modified and non-genetically modified soybeans. Results from 365 soybean fields indicated that 1998 yields from the genetically modified soybeans were slightly lower than from the conventional varieties, but so were the costs.

Other Aggregate Studies

Two previous studies have been conducted analyzing the aggregate impacts of the introduction of Roundup Ready soybeans. Auburn University researchers analyzed the impact of herbicide tolerant soybeans in 1997, using yield and cost change assumptions based on USDA survey data. Overall U.S. growers gains were estimated at $800 million. Table 19 shows assumptions used in the analysis and results of U.S. farmer surplus by region. The authors note that the results are extremely sensitive to the values chosen for certain variables, the supply elasticity in particular. Farmer surplus estimates are also presented using a different assumed elasticity, which reduces the gain to $126 million. Further, it should be noted that the assumptions on yield and cost changes are differences between adopters and non-adopters of the technology from the USDA survey data, as discussed above. These observed differences may be due to factors other than the adoption of herbicide tolerant varieties.

Monsanto cites the results of the Auburn study, concluding that Roundup Ready soybeans have reduced herbicide costs for U.S. soybean farmers by almost $700 million. However, this statement mischaracterizes the result of the analysis by attributing the impact to reductions in herbicide costs alone, while the correct interpretation of the Auburn study's result is that farm level impact estimates are based on a combination of yield and cost impacts.

Researchers at Iowa State University have also modeled the impact of adoption of Roundup Ready varieties. Several scenarios were considered, including varying assumptions about adoption patterns, market structure for the technology and yield changes. In the scenario most closely depicting what would be expected for crop year 1999-2000, an adoption rate of 55% for the U.S. is assumed, growers are assumed to realize a $20/ha cost savings and no yield advantage or disadvantage using Roundup Ready varieties. Change in producers surplus under this scenario was estimated at $156 million.

Herbicide Treatments

The USDA surveys of herbicide use in soybeans summarizes estimates by state for individual active ingredients that show the estimated percent of acres treated and the average number of applications per treated acre. Estimates of the total number of acres treated and average number of treatments per treated acre.

By multiplying the number of treated acres by the average number of applications per treated acre for each active ingredient, estimates are made of the total number of "application-acres" for each active ingredient. An application-acre is the number of different active ingredients applied per acre times the number of repeat applications. It is different than number of treatments, or passes over the field, in that a single treatment containing two ingredients is counted as two acre-treatments, as is two treatments containing a single ingredient.

A reduction in the number of application-acres reflects a reduction in the number of active ingredients used and/or the number of treatments. As can be seen, in 1995, on average, most active ingredients were used just slightly more than once per treated acre. In 1998, most active ingredients were used just once per treated acre. As can be seen, glyphosate was used on average 1.3 times per treated acre in 1998, which may represent a combination of a pre-plant burndown treatment with a postemergence spray, or two postemergence sprays on some acreage. Overall, the number of herbicide "application acres" totaled 12.5 million fewer in 1998 in comparison with 1995 in the thirteen states. As can be seen, the overall total of herbicide application acres declined by 9% between 1995 and 1998. This reduction occurred even though the total number of soybean acres increased by 12% between 1995 and 1998.

Inflating this subtotal to represent all soybean acreage implies an annual reduction of 16 million herbicide treatments on U.S. soybean acreage between 1995 and 1998. In 1995, the average treated soybean acre received 2.7 herbicide active ingredient applications while in 1998, the average treated acre received 2.2 applications, representing a 19% reduction.

A 1997 survey analysis by USDA's Economic Research Service (ERS) reported that there were statistically significant reductions in herbicide acre treatments in three regions by growers of herbicide tolerant soybeans in comparison with other growers [151]. For two regions, the ERS study reported a higher number of herbicide acre treatments by the herbicide tolerant soybean adopters, but these results were not statistically significant. It is believed that the use of glyphosate as a postemergence spray was the primary cause of the decline in usage of other active ingredients. In many cases, one glyphosate application substituted for two or more active ingredients, which would account for the decline in the number of herbicide application acres in U.S. soybeans between 1995 and 1998.

Herbicide Use Amounts

The USDA surveys of herbicide use include estimates of the total pounds of individual active ingredients used on soybeans by state. The USDA reports include totals by active ingredient, including usage during the growing season as well as preplant "burndown" applications. Table 22 summarizes the estimates of total pounds of herbicide active ingredients applied to soybeans in the thirteen states included in the USDA surveys for 1995 and 1998. As can be seen, the aggregate pounds applied went up by 14% (from 52 to 59 million pounds). One factor that led to the increased poundage was an increase in soybean acreage of 12% between 1995 and 1998. The average per acre use rate rose between 1995 and 1998. In 1995, there were 51.5 million soybean acres in the thirteen states, of which 97% were treated with herbicides, implying an average treatment rate of 1.04 lbs. In 1998, there were 57.8 million soybean acres in the thirteen states, of which 95% were treated with herbicides, implying an average application rate of 1.08 lbs. per acre.

Figure 7 shows the overall trend in herbicide use in soybeans in terms of millions of pounds per year from 1964 to 1998 (this figure is an extension of Figure 6). As can be seen, the overall poundage of herbicides went up between 1995 to 1998 – largely due to the increase in soybean acreage and to an increase in the rate per treated acre. It is not possible to determine whether the increase in herbicide use amounts occurred on the Roundup-Ready acreage or the non-Roundup-Ready acreage. For example, increases in the per acre treatment amount was recorded for 10 active ingredients in addition to glyphosate.

As can be seen, the average rate for glyphosate increased from 0.61 to 0.92 lbs. AI per acre, as the herbicide was used increasingly to control weeds during the growing season in addition to burndown or spot treatments. Many of the herbicides that glyphosate displaced are used at lower rates. The rates of alternative herbicides are lower than glyphosate's: e.g., acifluorfen (0.24 lb/acre), chlorimuron (0.02 lb/acre), imazethapyr (0.04 lb/acre), etc. However, most of the alternatives to glyphosate are used in combinations, increasing the total number of pounds per acre. For example, a common treatment is pendimethalin and imazethapyr with an overall rate of 1.09 lbs. per acre. There is an absence of information regarding the distribution of the exact programs used by soybean growers in 1995 and 1998. Undoubtedly, some growers in a Roundup Ready program made two postemergence applications while others may have made one postemergence in combination with a preemergence trifluralin or pendimethalin application.

Monsanto commissioned two studies (for 1996 and 1997) that compare the in-season use of herbicides in Roundup-Ready soybean fields with in-season herbicide use in non-Roundup-Ready fields. As can be seen, in-season herbicide use, as measured in terms of pounds of active ingredient used in-season per acre, was lower in the Roundup-Ready fields in all regions for both years. Monsanto reports that 77% and 70% of those surveyed said that they only had to apply glyphosate once to control weeds for the entire growing season in 1996 and 1997, respectively. Twenty-two to 29% made two applications with glyphosate and 1% made three or more trips.

One analyst has claimed that farmers growing Roundup Ready soybeans use two to five times more herbicides measured in pounds applied per acre. However, this statement is based merely on a comparison of the rate of glyphosate with the rates of other postemergence herbicides with no survey data on actual usage.

Other Impacts

As noted above, the use of glyphosate in soybean fields imposes no minimum interval before a different crop can be planted in a rotation. Since many of the alternative commonly-used herbicides do have minimum intervals, it is likely that at some point additional crops will be grown in rotation with soybeans. However, it is still too early to tell whether crop rotations have (or will) shift because of the greater flexibility in planting a different crop following soybeans.

Weed populations resistant to many commonly-used soybean herbicides had developed to a significant level in the mid 1990's. As noted above, herbicide-resistant kochia and waterhemp infested a sizable portion of the soybean growing regions of Missouri and Kansas. Following the introduction of Roundup Ready soybeans, soybean growers had an effective alternative to control these resistant weed populations. As a result, many more acres of soybeans were planted.

TABLE : Herbicide Use on U.S. Soybean Acreage: (% Acreage Treated, 1995-1998)

Active Ingredient

	1995	1996	1997	1998
Acifluorfen	12	11	12	7
Alachlor	4	5	3	2
Bentazon	12	11	11	7
Chlorimuron	16	14	13	12
Fluazifop	10	7	7	5
Glyphosate	20	25	28	46
Imazaquin	15	15	13	8
Imazethapyr	44	43	38	17
Metolachlor	7	5	7	4
Metribuzin	11	9	10	6
Pendimethalin	26	27	25	18
Sethoxydim	7	9	7	5
Trifluralin	20	22	21	16
Thifensulfuron	12	10	9	5

Source : Selected major herbicides only. Includes preplant and in-season use.

TABLE : Herbicide Prices, 1990-1998 ($/LB/AI)

			Active		Ingredient				
	1990	1991	1992	1993	1994	1995	1996	1997	1998
Acifluorfen	30	29	29	29	30	30	31	32	34
Alachlor	6	6	6	6	6	7	7	6	6
Bentazon	15	16	16	16	17	17	18	18	18
Chlorimuron	1126	1164	1137	1123	1123	1210	1220	620	722
Fluazifop	94	97	59	59	59	66	66	62	63
Glyphosate	21	18	15	16	16	17	18	18	14
Imazethapyr	287	284	290	299	310	318	329	340	200
Metribuzin	30	32	32	34	36	40	35	36	27
Pendimethalin	7	9	9	8	8	8	8	9	8
Sethoxydim	70	77	60	59	59	59	56	56	51

TABLE : Aggregate Herbicide Expenditures: Soybeans (13 States)

$/LB/AI₁ 000 LB AI₂ 000$/YR₃ Active Ingredient	1995	1998	1995	1998	1995	1998
2,4-D	3	4	2,248	1,910	7,643	7,411
Acifluorfen	30	34	1,362	1,041	40,656	34,946
Alachlor	7	6	2,749	746	18,785	4,590
Bentazon	17	18	4,241	1,839	73,115	33,966
Chlorimuron	1,210	722	151	140	182,687	101,125
Clethodim	120	97	200	288	23,977	27,825
Clomazone	22	20	605	1,281	13,358	25,569
Cloransulam	-	491	0	2	0	982
Dimethenamid	13	13	286	438	3,850	5,895
Ethalfluralin	11	11	0	0	0	0
Fenoxaprop	114	102	316	315	36,011	32,250
Fluazifop	66	63	359	153	23,522	9,581
Flumetsulam	188	188	12	36	2,257	6,771
Fomesafen	33	46	552	830	18,183	38,495
Glyphosate	17	14	6,313	24,944	105,238	358,695
Imazamox	-	600	0	113	0	67,800
Imazaquin	160	158	716	360	114,718	56,772
Imazethapyr	318	200	1,332	401	423,829	80,200
Lactofen	60	66	145	43	8,651	2,851
Linuron	21	22	217	22	4,659	495
Metolachlor	9	10	6,745	3,673	57,400	37,722
Metribuzin	40	27	1,259	674	50,612	18,269
Paraquat	14	15	365	264	4,979	3,889
Pendimethalin	8	8	12,897	10,280	107,690	85,632
Quizalofop	154	156	134	59	20,592	9,178
Sethoxydim	59	51	646	559	37,849	28,685
Sulfentrazone	-	78	0	250	0	19,500
Thifensulfuron	2,249	2,464	19	4	42,740	9,855
Trifluralin	9	9	8,125	8,633	69,117	76,963
TOTAL	51,994	59,298	1,49	2,115	1,18	5,913

1. From

2. From Includes the states: AR, IL, IN, IA, KY, LA, MN, MS, MO, NE, NC, OH, TN, which account for 80 % of U.S. soybean acreage

3. Calculated by multiplication Includes preplant and in-season use.

TABLE : Yields Comparison of Herbicide-tolerant Soybeans to Conventional Varieties by Production Region (% difference)

Region	1996	1997	1998
Heartland	-2.4	13.6 **	4.4
Mississippi Portal	2.8	-6.2	-3.6
Northern Crescent	id	-4.6	4.8
Prairie Gateway	id	21.0 **	24.2
Southern Seaboard	id	13.3 *	21.4 *
Eastern Uplands	id	id	-8.0

Herbicide tolerant varieties include both Roundup Ready and STS varieties.
id-insufficient data to estimate
* significantly different from all other at the 10 percent level.
** significantly different from all other at the 5 percent level.
Source: ERS Study

TABLE : Herbicide Applications to Soybeans (13 States)

	Percent Acres Treated		# of Applications Per Treated Acre		Total Application-Acres (000)	
Active Ingredient	**1995**	**1998**	**1995**	**1998**	**1995**	**1998**
2,4-D	10	8	1.00	1.00	5,077	4,613
Acifluorfen	11	7	1.10	1.02	6,420	4,376
Alachlor	3	1	1.01	1.04	1,501	490
Bentazon	12	5	1.03	1.00	6,462	2,672
Chlorimuron	15	13	1.08	1.02	8,600	7,908
Clethodim	4	4	1.01	1.01	2,212	2,115
Clomazone	2	3	1.00	1.00	1,081	1,809
Cloransulam	-	<1	-	1.00	-	128
Dimethenamid	1	1	1.00	1.00	284	415
Fenoxaprop	5	4	1.08	1.03	2,842	2,467
Fluazifop	10	5	1.09	1.04	5,657	2,873
Flumetsulam	<1	1	1.00	1.00	236	663
Fomesafen	4	6	1.05	1.02	1,957	3,316
Glyphosate	20	47	1.01	1.30	10,478	35,163
Imazamox	-	7	-	1.00	-	3,990
Imazaquin	15	7	1.10	1.01	8,346	4,231
Imazethapyr	44	16	1.00	1.00	22,522	9,413

Lactofen	4	1	1.00	1.00	1,840	539
Linuron	1	<1	1.00	1.00	451	44
Metolachlor	7	4	1.02	1.00	3,520	2,098
Metribuzin	9	6	1.04	1.00	5,080	3,184
Paraquat	1	1	1.12	1.00	645	531
Pendimethalin	26	17	1.10	1.00	14,810	9,791
Quizalofop	5	2	1.03	1.00	2,582	1,159
Sethoxydim	7	4	1.00	1.00	3,601	2,580
Sulfentrazone	-	4	-	1.00	-	2,087
Thifensulfuron	12	4	1.04	1.00	6,701	2,540
Trifluralin	19	16	1.01	1.00	10,069	9,256
Total		132,974			120,445	

Source: Derived from

SUMMARY AND CONCLUSIONS

Soybeans are an extremely important and valuable crop in the United States. Controlling weeds is the major pest problem that soybean growers must plan for every year. Before the introduction of herbicides, soybean growers made many trips across fields with cultivators, and yet, the nation lost about 17% of national production to weeds because of limited effectiveness of the mechanical operations.

The use of herbicides in soybeans for weed control is one of the most important technological changes that has made possible consistent high yields for U.S. soybeans. Since herbicides are more effective than non-chemical controls, soybean yields increased dramatically as growers switched to them and reduced tillage operations.

Soybean cultivars have been improved genetically through crossbreeding for many decades. All currently grown U.S. soybean cultivars are the result of genetically improving cultivars imported from Asia. As important as herbicides are for soybean production, genetically improving soybean cultivars to tolerate applications of these chemicals has not been a focus of conventional crop breeding. As a result, chemicals have been selected for herbicidal use in soybeans based on screening compounds that kill weeds without causing major damage to soybeans. Although soybean plants detoxify these chemicals, several of the most widely used herbicides in soybeans can cause serious injury to the crop under certain environmental conditions.

As a result of changing one enzyme out of the approximately 10,000 enzymes in a soybean plant, scientists created a transgenic plant that tolerates the use of the broad-spectrum herbicide glyphosate. The plant is called transgenic because the gene that conferred glyphosate tolerance was taken from a soil bacterium and inserted into soybean DNA by a non-sexual method. Glyphosate tolerance in the transformed plant is based on the insertion of DNA for a different form of an enzyme already present in the soybean plant. The difference in the enzyme from the bacterium means that glyphosate does not interrupt the functioning of the enzyme. U.S. growers have planted the transgenic herbicide tolerant soybean varieties on a large percentage of U.S. soybean acreage primarily because of the following factors:

- Less complicated weed control
- Broader spectrum weed control
- Less crop injury
- More flexibility in timing treatments
- Less concern of carryover to rotational crops

In addition to these factors, the herbicide tolerant seed and associated herbicides were priced competitively with existing herbicide programs for soybeans. Following the introduction of the herbicide tolerant transgenic soybeans, the prices of alternative herbicide treatments were lowered significantly. Overall, U.S. growers saved approximately $220 million in 1998 due to lower herbicide costs.

The broad spectrum of weeds controlled by glyphosate means that soybean growers no longer need to make as many multiple applications with combinations of herbicides. As a result of the widespread substitution of one active ingredient application for an application that might include three or four active ingredients, U.S. soybean growers made 16 million fewer herbicide active ingredient applications in 1998 in comparison with 1995. A 19% reduction occurred in the number of herbicide active ingredient applications made to the average treated soybean acre.

Bibliography

Adelberg, Edward A.: *Papers on Bacterial Genetics*, Boston, Brown and Company, 1960.

Ausubel, F.M.: *Current Protocols in Molecular Biology*, New York, John Wiley and Sons, 1989.

Bailey, Ronald: *Liberation Biology: The Scientific and Moral Case for the Biotech Revolution*, Amherst, Prometheus Books, 2005.

Barnes, N.: *Biology*, New York, Worth Publishers, 1989.

Barnum, Susan R.: *Biotechnology: An Introduction*, Belmont, Thomson/Brooks/Cole, 2005.

Bernauer, Thomas: *Genes, Trade, and Regulation: The Seeds of Conflict in Food Biotechnology*, Princeton, Princeton University. 2003.

Brandwein, P.F.: *Sourcebook for the Biological Sciences*, San Diego, Harcourt Brace JOvanovich, 1986.

Brooks, Rodney A.: *Flesh and Machines: How Robots Will Change Us*, New York, Pantheon Books, 2002.

Chrispeels, Maarten : *Plants, Genes and Crop Biotechnology*, Sudbury MA, Jones and Barlett Publishers, 2003.

Conko, Gregory: *The Frankenfood Myth: How Protest and Politics Threaten the Biotech Revolution*, New York, Praeger, 2004.

Cox, Earl D. : *Beyond Humanity: Cyber Evolution and Future Minds*, Roackland, Charles River Media, 1996.

David Sadava: *Plants, Genes and Crop Biotechnology*, Sudbury MA, Jones and Barlett Publishers, 2003.

DeGregori, Thomas R.: *Bountiful Harvest: Technology, Food Safety and the Environment*, Washington DC, Cato Institute, 2003.

Dixon, Dougal: *After Man-A Zoology of the Future*, New York, St. Martin's Press, 1981.

Dodds, John H.: *Plant Genetic Engineering*, New York, Cambridge University Press, 1985.

Drexler, K. Eric: *Engines of Creation: The Coming Era of Nanotechnology*, New York, Anchor Books, 1986.

Fedoroff, Nina Y.: *Mendel in the Kitchen: A Scientist's View of Genetically Modified Foods*, Washington, Joseph Henry Press, 2004.

Fransman M, Junne G, Roobeek A: *The Biotechnology Revolution?*, Oxford, Blackwell, 1995.

Friedberg, E.C.: *DNA Repair*, New York, WH Freeman and Company, 1985.

Fukuyama F.: *Our Posthuman Future: Consequences of the Biotechnology Revolution*, NewYork, Farrar, Straus and Giroux, 2002.

Fumento, Michael: *Bioevolution: How Biotechnology is Changing Our World*, San Francisco, Encounter Books, 2003.

Gaskell G: *Biotechnology-the Making of a Global Controversy*, Cambridge, Cambridge University Press, 2002.

George S. Paul: *Beyond Humanity: Cyber Evolution and Future Minds*, Roackland, Charles River Media, 1996.

Goodman, D.C.: *From Farming to Biotechnology: A Theory of Agro-industrial Development*, Oxford, Blackwell, 1987.

Goodsell, David S.: *Bionanotechnology: Lessons From Nature*, Hoboken, Wiley-Liss, 2004.

King, R.B., Mihony, G., and Sheldon, J.K.: *Practical Environmental Bioremediation*, Lewis Publishers, Boca Raton, 1992.

Kurzweil, Ray: *The Age of Spiritual Machines*, New York, Penguin Books, 1999.

Metting, Jr., F.B.: *Soil Microbial Ecology: Applications in Agricultural and Environmental Management*, Marcel Dekker, New York, 1992.

Miller, Henry I. : *The Frankenfood Myth: How Protest and Politics Threaten the Biotech Revolution*, New York, Praeger, 2004.

Moravec, Hans: *Mind Children: The Future of Robot and Human Intelligence*, Cambridge, Harvard University Press, 1988.

Morholt, E. : *Sourcebook for the Biological Sciences*, San Diego, Harcourt Brace JOvanovich, 1986.

Murray, David: *Seeds of Concern: The Genetic Manipulation of Plants*, Sydney, University of New South Wales, 2003.

Nancy M. Brown: *Mendel in the Kitchen: A Scientist's View of Genetically Modified Foods*, Washington, Joseph Henry Press, 2004.

Old, R.W. : *Principles of Gene Manipulation*, London, Blackwell Scientific Publications, 1989.

Parens, Erik: *Enhancing Human Traits: Ethical and Social Implications*, Washington, D.C.: Georgetown University Press, 1998.

Poindexter, J.S.: *Strategies and Mechanisms for Field Research in Environmental Bioremediation*, American Academy of Microbiology, Washington, D.C., 1994.

Postman, Neil: *Technopoly: The Surrender of Culture to Technology*, New York, Vintage Books, 1992.

Primrose, S.B.: *Principles of Gene Manipulation*, London, Blackwell Scientific Publications, 1989.

Retzer, W.J.: *Biotechnology Workbook*, Englewood Cliffs, Prentice Hall, 1991.

Rifkin, Jeremy: *The Biotech Century*, New York, Penguin Putnam, 1998

Sayler, G.S., Fox, R., and Blackburn, J.W.: *Environmental Biotechnology for Waste Treatment*, Plenum Press, New York, 1991.

Shetty, Kalidas: *Food Biotechnology*, New York, Dekker/CRC Press, 2005.

Slater, R.J.: *Experiments in Molecular Biology*, Clifton, Humana Press, 1986.

Smith, John E.: *Biotechnology*, Cambridge, Cambridge University Press, 2004.

Switzer, R.L.: *Experimental Biochemistry*, New York, W.H. Freeman and Company, 1977.

Towle, Albert: *Modern Biology*, Austin, Holt, Rinehart and Winston, 1988.

Walden, Richard: *Genetic Transformation in Plants*, England, Open University Press, 1988.

Winston, Mark L.: *Travels in the Genetically Modified Zone*, Cambridge, Harvard University, 2004.

Index